Genetic Design Automation

Hasan Baig • Jan Madsen

Genetic Design Automation

A Practical Approach for the Analysis,
Verification and Synthesis of Genetic
Logic Circuits

 Springer

Hasan Baig
Center for Quantitative Medicine,
UConn Health
University of Connecticut
Connecticut, CT, USA

Jan Madsen
Department of Applied Mathematics
and Computer Science
Technical University of Denmark (DTU)
Lyngby, Denmark

ISBN 978-3-030-52357-2   ISBN 978-3-030-52355-8   (eBook)
https://doi.org/10.1007/978-3-030-52355-8

This Springer imprint is published by the registered company Springer Nature Switzerland AG
The registered company address is: Gewerbestrasse 11, 6330 Cham, Switzerland

To our families.

Preface

In 1961, François Jakob and Jacques Monod, two pioneering French molecular biologists, speculated that gene regulating elements could in principle be combined to produce custom systems for control of genetic activity. Gene regulation is an essential mechanism for cell metabolism, which allows the cell to increase or decrease the production of specific proteins. As the production of one protein may be used to regulate the production of other proteins, cells contain sophisticated and often complex gene-regulatory networks to respond to various conditions, such as environmental stimuli, adapt to new food sources, or trigger developmental pathways. With the advent of DNA sequencing, it became possible for researchers to identify and classify gene regulating elements, effectively creating libraries of such elements or components. This allowed researchers to combine these components into novel configurations to modify existing properties or to create or synthesize entirely new ones. Synthetic biology had then emerged as a new research field.

In 2000, two important contributions pointed towards the capabilities of synthetic biology. James Collin's laboratory demonstrated the possibility to create a memory function into bacterial host cells. They created a network of two genes, each repressing the activity of the other gene. This simple network had two states of stability, which could be switched between by applying either a sugar-like drug or an antibiotic, in much the same way that a set–reset latch works in a classical computer memory circuit. The other contribution was the demonstration of an oscillator circuit by the laboratory of Michael Elowitz and Stanislas Liebler. In this case, the network consisted of three genes repressing each other in sequence, in much the same way as a ring oscillator can be constructed by three inverter circuits connected in a loop using classical computer circuits. From these two genetic circuits, a large number of other genetic logic circuits followed, some connected to sensing systems of the cell, making them react to the cell's environment in real-time. Such systems have been explored to address specific purposes in fields like energy, environment, and medicine and have created a new branch of synthetic biology, focusing on synthetic genetic logic circuits and sensors.

Still today, new genetic circuits are to a large extent formed in an experimental bottom-up approach, where the genetic circuit is assembled from libraries of characterized genetic components to match an intended behavior and tested in the laboratory. This approach resembles how microelectronics circuits are designed by connecting library components to form more complex circuits and eventually full computer systems. The success of microelectronics is due to the successful creation of abstract models that allow a top-down synthesis approach, where the intended behavior is expressed in a high-level language, validated by simulation and then automatically synthesized into the most efficient circuit design. When the complexity of the genetic logic circuits grows, a top-down approach similar to the electronic design automation (EDA) of microelectronic circuits will be essential.

This book is intended for everyone, computer scientists, engineers, and biologists, and provides a foundation for a complete tool flow for genetic design automation (GDA), similar to that of logic synthesis for electronic logic circuits. It provides enough background knowledge to computer scientists or engineers who usually do not have any background in biology and are willing to start off their career in this domain.

Our aim of this book is twofold: (1) to provide an introduction to the fascinating technology of genetic logic circuits, for researchers working in the field of electronic design automation field, and (2) to provide a textbook for students to learn about genetic design automation by providing exercises, student projects, as well as a hands-on introduction to GDA tool development using the graphical programming language, LabVIEW. With the help of LabVIEW and some small sample projects, readers are guided to start developing GDA tools rapidly. This part of the book is also helpful for biologists who usually find programming difficult to grasp and are reluctant to dive into computer software development. In particular, we present tool solutions for

- Genetic circuit simulation allowing virtual experimentation in real-time.
- Logic validation through timing and threshold analysis of genetic logic circuits.
- Logic optimization and technology mapping of genetic logic circuits.

The proposed tool flow is validated against a set of genetic logic gates produced and laboratory tested by a group of researchers at MIT and Boston University.

This book is a result of several years of research conducted at the Technical University of Denmark (DTU), Denmark, and continued first at Habib University, Pakistan, and later at University of Connecticut, USA. This work has been greatly influenced by many discussions with colleagues and researchers around the world. For this, we would like to thank in particular Chris Myers from University of Utah, Alec Nielsen from MIT, Nicholas Roehner, Doug Densmore, Sonya Iverson, Ernst Christoph Oberortner, and Swapnil Bhatia from Boston University, and Christopher Workman from DTU Bioengineering. Furthermore, we would like to thank our students Sune Mølgaard Laursen and Jakob Jakobsen Boysen (from DTU) for their early contributions to the optimization of genetic logic circuits; Sanaullah Rao (from Chosun University) for enhancing simulation tool; and Mudasir Shaikh, M. Ali Bhutto, Mukesh Kumar, M. Abdullah Siddiqui, and Adil Ali khan (all from Habib University) for contributing to the improvements of the technology mapping tool. We also acknowledge Prof. Jeong A lee (Chosun University, Korea) for supporting her student to continue enhancing the simulation tool and Dr. Faisal Khan (Averos Inc., Pakistan) for accommodating students in his lab for wet lab experimentation. For funding, we thank DTU for providing the PhD scholarship on which this book is based. Finally, we would like to thank Springer Nature and especially Charles Glaser and his team for publishing this work.

Connecticut, CT, USA   Hasan Baig
Lyngby, Denmark   Jan Madsen
April 2020

Contents

Acronyms

ABS	Activator-binding site
CAD	Computer aided design
cAMP	Cyclic adenosine monophosphate
CAP	Catabolite activator protein
CellML	Cell markup language
DAG	Directed acyclic graph
DNA	Deoxyribonucleic acid
EDA	Electronic design automation
GDA	Genetic design automation
GFP	Green fluorescent protein
GRN	Gene-regulatory network
HDL	Hardware description language
IPTG	Isopropyl $\beta - D - 1 - thiogalactopyranoside$
IUPAC	International Union of Pure and Applied Chemistry
LabVIEW	Laboratory Virtual Instrument Engineering Workbench
LED	Light emitting diode (*LabVIEW indicator*)
MathML	Mathematical markup language
MOSFET	Metal oxide semiconductor field effect transistor
mRNA	Messenger ribonucleic acid
ODE	Ordinary differential equation
PoPs	Polymerase per second
POS	Product of sum
QSG	Quick start guide
RBS	Ribosome binding site
RiPS	Ribosomes per second
RNA	Ribonucleic acid
RNAP	RNA polymerase
SBML	Systems biology markup language
SBOL	Synthetic biology ontology language
SBOLv	SBOL visual
SOP	Sum of product

SSA	Stochastic simulation algorithm
VI	Virtual instrument
XML	eXtensible markup language
YFP	Yellow fluorescent protein

Part I
Introduction

In this part, you will be given an introduction to a concept of computations in living cells using genetic logic circuits. You will also be given a brief introduction of molecular biology. This section consists of two chapters.

Chapter 1
Introduction

An advancement in the understanding of cellular processes and DNA synthesis methods suggests that the living cells can be viewed as a *programmable* matter. With this revolutionary finding, logical computations can be performed inside a living cell through a group of biological components, collectively called *genetic circuits*. A genetic circuit represents a gene-regulatory network (GRN), which is composed of small genetic components. These components interact with the external signals (like temperature, light, proteins, etc.) to control the behavior of a living cell.

Genetic circuits are a key application of *synthetic biology* which is an emerging engineering discipline to program cell behaviors as easy as computers are programmed. Synthetic biology is defined by syntheticbiology.org as:

(a) the design and construction of new biological parts, devices, and systems and (b) the re-design of existing natural biological systems for useful purposes.

Biologists are interested in synthetic biology because it provides a viewpoint to analyze, understand, design, and ultimately build a biological system. Engineers, on the other hand, are attracted toward synthetic biology because the living world has the abundant mechanisms for controlling life behavior and processing information.

1.1 Why Computations in Cells?

There are numerous complex computations a living cell performs on the continuous environmental signals they encounter. The natural biological systems can be engineered to perform sophisticated computations in living cells. Biologists and engineers are working together on synthetic biology [1] to design new and useful biological systems. The synthetic biological systems performing digital [2] and analog [3] computations have already been implemented.

© Springer Nature Switzerland AG 2020
H. Baig, J. Madsen, *Genetic Design Automation*,
https://doi.org/10.1007/978-3-030-52355-8_1

The artificial computation in living cells will revolutionize the industry of medicine and biotechnology. The aim of performing synthetic computations in living cell is to develop genetic devices to address real-world problems. To name a few, these problems include the development of genetic systems to detect and destroy cancer cells [4]; production of liquid biofuels to address the global energy and environmental problems [5]; consuming toxic wastes to avoid environmental pollution; and the production of drugs to treat health problems like Malaria [6], or the deadly pandemic virus, COVID-19.

1.2   State-of-the-Art

Similar to *electronic design automation* (EDA) processes which dramatically enhanced the design, verification, validation, and production of electronic circuits, researchers have started to work on the development of *genetic design automation* (GDA) tools [7] to automate the design, test, verification, and synthesis processes of genetic circuits prior to their validation in laboratory. There are several GDA tools (see Sect. 2.4) which allow synthetic biologists to design genetic circuits at a high level of abstraction with the focus on a desired function, rather than exact genetic components used to achieve this functionality. By encoding standardized data, genetic constraints, and the components library in GDA tools, the process of genetic circuit construction and analysis has been automated. This not only has reduced the lengthy design process and iterative tests for constructing complex genetic circuits, but has also promoted the reuse of experimentally tested genetic components.

The modern trend to analyze genetic circuits is to perform *in-silico* (in computer) analysis either by solving ordinary differential equations (ODEs) or by performing stochastic simulations, with the aim to reduce the number of required in-vitro (in laboratory) experiments. In order to perform these analyses in a computer, models of biological systems must be represented in a standard computerized format. Several different methods have been proposed to represent and analyze genetic systems [8]. Among these methods, the most widely used standards to represent the *behavior* and the *structure* of a genetic model are the Systems Biology Markup Language (SBML) [9], and the Synthetic Biology Open Language (SBOL) [10], respectively.

SBML allows users to define the behavior of a circuit by specifying the species of a genetic network and how they interact with each other through chemical kinetics. SBOL, however, is used to illustrate genetic designs graphically with the help of standardized vocabulary of schematic glyphs (SBOL Visual) as well as standardized digital data (SBOL Data). More information on *standards* can be found in Sect. 2.3.

1.3 Motivation

Synthetic biology not only aims to play with natural biological systems but also to construct artificial complex systems from the library of well-characterized biological components, in a similar way as electronic circuits are designed and constructed. While comparison with electronic circuits is useful, there are several important challenges which make the design of genetic circuits more challenging. For instance, genetic components, in contrast to electronic components, are not physically separated from each other. This not only makes the reuse of genetic components in the same system more difficult, but also increases the crosstalk with the neighboring circuit components. Also, the electronic logic gates are composed of transistors which have well defined and uniform threshold voltage levels that categorize the logic levels 0 and 1. However, in genetic circuits, each genetic gate is composed of different genetic components which results in the different threshold concentration values. Additionally, in comparison to electronic circuits which have the same physical quantities as input and output signals, the genetic circuits have different species at the input and other at the output, which makes the genetic modules integration more difficult.

As electronic engineers develop circuits using electronic logic gates (such as AND, NAND, and NOT gates), genetic engineers use biological equivalents of these components to control the function of a cell [2, 11]. The field of genetic circuit design is still immature and only small circuits, containing limited number of genes, can be constructed in the laboratory. However, genetic circuits can be designed from a very large number of genetic parts [12] creating a large space of possible solutions even for circuits of limited complexity.

The current practice is to design such circuits directly in the laboratory, through trial and error, which is a time consuming and costly process, as thousands of circuits may have to be tested in order to find a few that works. Due to this, the process of design and implementation of genetic circuits remain very slow. To address these challenges, it is necessary to improve computer aided design (CAD) tools to speed up the design and analyses procedures of genetic circuits. In particular, it is necessary to develop tools which allow genetic design engineers to capture and analyze the stochastic behavior of biological systems dynamically in a way that sounds natural to them.

1.4 Scope of This Book

An electronic design engineer would never fabricate a circuit on silicon prior to its functional validation and behavioral analysis. Similarly the most important phase in GDA is the simulation and *in-silico* (in computer) analysis of genetic circuit models to increase the chances that the system would work *in-vivo* (in living organism) correctly. There are plenty of tools developed to simulate the behavior of genetic circuits [13]. These tools, however, lack some important and useful features which can not only increase the designer's productivity but also help them design genetic circuit models more effectively. Out of many challenges in the field of GDA, some of the challenges, listed below, have been addressed in this book. We believe that addressing the following challenges will not only increase the productivity of genetic design engineer but will also increase the reliability and robustness of genetic circuit models.

1.4.1 Virtual Experimentation

First, it would be very helpful for biologist or design engineers to have a tool which allow them to perform laboratory experiments virtually in-silico. This corresponds to an experimental environment where a user can trigger the concentrations of input species or change the parameter values (for example, increasing temperature) at any instant of time and observe their live effects on the model's behavior. For in-silico analyses, the standard way to capture the instantaneous, discontinuous state change in the model is by defining *events* (see Sect. 2.3 for more details). For example, events (shown as green-boxes in Fig. 2.6b) are used to trigger the concentration of input species to a certain level, at a specific point in time, and to observe the effects on the concentration of output species. A single event can be used to represent only one instance of triggering the concentration to a certain level at a specific time. Since events are predefined, they cannot be changed during run-time, which means that the output of a genetic circuit can be observed only for defined events. In order to observe the output, the different set of input conditions, i.e., when to change what input to which level, must be defined in each event. Even for moderate sized genetic circuits, capturing all combinations of inputs and concentration levels may require a very large number of events to be defined and simulated. To the best of our knowledge, there exist no tools that allow users to trigger/change input species on the fly during the simulation, effectively creating a *virtual lab*.

1.4.2 Timing and Threshold Analysis

In contrast to EDA tools which allow a user to perform timing analyses, to the extent of our knowledge, no GDA tool allows a user to perform timings and threshold value analyses for genetic gates/circuits. Electronic design engineers do not need to estimate the threshold value for each electronic circuit because these values are well defined and holds valid for all electronic logic gates. However, this is not the case for genetic gates where each of them are composed of different components and have different input and output molecular identities. Therefore each genetic gate may have different input threshold values and thus exhibit different timing behaviors. It is therefore necessary to have such a tool which should assist a user in identifying the correct input threshold concentration required to trigger the circuit's output along with the estimation of propagation delays. It may also help a user to perform in-vitro experimentation quickly by applying the estimated threshold concentration values at input (rather than following *trial-and-error* approach) and expect the circuit's output to be triggered approximately within the time estimated as a propagation delay.

Similar to electronic circuits where timing analysis is a vital design characteristic, the timing analysis may also become an essential design characteristic in genetic circuits. It is therefore very important to have such analyses *in-silico* prior to the circuit's implementation *in-vivo*.

1.4.3 Automatic Logic Validation

It is also interesting to automatically validate if the behavior of a genetic circuit complies with the design rules. For example, the behavior of a genetic AND gate can be validated by applying all the possible input combinations and determine if the circuit's response obeys the AND Boolean logic. It might be easier to analyze the logical behavior of small circuits by just looking at response curves, but it may become a cumbersome task if the behavior is to be validated manually for complex genetic circuits. Therefore, an automated approach for analyzing the logic in genetic circuits will be helpful.

1.4.4 Effortless Circuit Designing

One of the several challenges in making the design process of genetic gates easier and user-friendly is to let the designers construct genetic circuits at a very high level of abstraction. Recently a tool, named Cello, is developed [14] which allows users to program genetic circuits as easy as electronic circuits are designed through *hardware description language* (HDL). Cello provides user, specially *computer scientist*, a fairly high level of abstraction to develop genetic circuits without worrying about the underlying physics of genetic interactions. However, this still requires a *biologist* to learn programming principles and the syntax in which the design module should be written. To let the *biologists* design genetic circuits effortlessly without additional prerequisites, a further simple and straightforward mechanism should be developed.

1.5 Book Organization

This book is organized as follows. **Chapter 2** gives the information about genetic circuits. This chapter gives some basic knowledge of genetic terminologies and a brief overview of how genetic systems work. It also describes the *standards* in more detail and also give information about existing GDA tools.

Chapter 3 gives an overview of Virtual Experimentation using DVASim. It briefly describes the whole simulation flow with the help of an example circuit model. The details of each subsequent step in this flow are discussed in separate chapters.

In **Chap. 4**, the methodology of timing analysis of genetic logic circuits is presented. This chapter discusses the algorithm developed for analyzing the threshold value and propagation delay of a genetic circuit model. The experimental results are included to support the significance of timing analysis in genetic logic circuits.

Chapter 5 explains the methodology developed to analyze and verify the logical behavior, of a genetic circuit, from the stochastic simulation data. This chapter also contains the experimental results of logic analysis on different genetic circuit models and the performance evaluation of the algorithm.

The approach for synthesis and technology mapping of genetic circuits is provided in **Chap. 6**. This chapter begins with describing the algorithms developed for reducing the Boolean expression of genetic gates into an optimized form, followed by its synthesis into NOR-NOT form. Then the methodology of technology mapping, along with the discussion of how the genetic gates library is constructed from the data disclosed in [14], is presented. In the end, some experimental results on case study have been presented.

Chapter 7 introduces the audience to a graphical (G) programming language platform. After completing this chapter, you will have enough knowledge and confidence to start developing programs using G language.

Chapter 8 is a hands-on project which teaches you how to implement a very simple form of stochastic simulation algorithm using graphical programming language. The chapter is followed by a *challenge* to enhance the development skills further.

Chapter 9 is another hands-on project which gives you a brief introduction of SBML and teaches you how to develop an SBML parser using graphical programming language. This chapter also contains a *challenge* to further improve the development skills.

Figure 1.1 shows a very high-level diagram which describes how the methods and tools discussed in this book can be used. Having the SBML model of a genetic circuit in DVASim, users can perform ODE and stochastic simulations in an environment which gives them a feeling of being in the lab performing live experiments by interacting with the model during run-time. Similar to many EDA tools which allow hardware design engineers to perform timing analysis of electronic circuits, DVASim is the first tool which provides users an ability to perform the timing analysis of genetic circuits. Furthermore, the experimental data, generated from stochastic simulations, can be used to analyze the logical behavior of a genetic circuit. Another tool, *GeneTech*, takes a raw Boolean expression as an input and generates all the possible circuits (in the form of structure) to achieve a desired logic. GeneTech generates the output in the form of standard SBOL

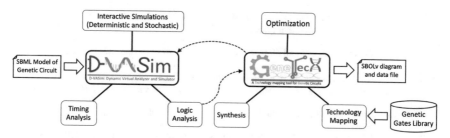

Fig. 1.1 The abstract diagram showing how the methods and tools discussed in this book can be used

data files as well as pictorial representation. The dotted line between GeneTech and the DVASim logic analyzer shows that the Boolean expression generated from the logic analyzer can also be used to obtain other possible circuits for the model being simulated. The circuits are generated using the genetic gates library [14]. The generated SBOL models of genetic circuit can be converted into SBML form using any SBOL-to-SBML conversion tool (like iBioSim [15]) and then can be analyzed back in DVASim again.

1.6 Resources

DVASim latest released version and the corresponding quick start guide (QSG) are available to download from http://bda.compute.dtu.dk/downloads/d-vasim/.

The latest released version of GeneTech and the QSG can be downloaded from http://bda.compute.dtu.dk/downloads/genetech/.

Both DVASim and GeneTech are open source tools and their source codes can be accessed at the following links:

DVASim: https://github.com/hasanbaig/D-VASim.

GeneTech: https://github.com/hasanbaig/GeneTech.

Section III resources are provided as a supplementary material with this book. Solutions for all exercises and project challenges are available separately for instructors.

1.7 Conventions Used in This Book

The following conventions appear in this book:

This icon denotes a tip or hint or an idea, which notifies you to advisory information or further clarification.

This icon denotes an alert, which notifies you to important information.

Bold   Bold text is used to indicate the steps to follow; option selection; built-in LabVIEW function/structure, procedure names, and definitions.

Bold Italic   ***Bold italic*** text is used to denote the file names and VI names.

Italic   *Italic* text is used to emphasize something or to denote the names of program variables.

References

1. A. Arkin, Setting the standard in synthetic biology. Nat. Biotechnol. **26**(7), 771–774 (2008)
2. J.J. Collins, T.S. Gardner, C.R. Cantor, Construction of a genetic toggle switch in Escherichia coli. Nature **403**(6767), 339–342 (2000)
3. R.W. Basu, S. Basu, The device physics of cellular logic gates, in *NSC-1: The First Workshop of Non-Silicon Computing*, vol. 158 (2002), pp. 39–41
4. J.C. Anderson, E.J. Clarke, A.P. Arkin, C.A. Voigt, Environmentally controlled invasion of cancer cells by engineered bacteria. J. Mol. Biol. **355**(4), 619–627 (2006)
5. S. Atsumi, J.C. Liao, Metabolic engineering for advanced biofuels production from Escherichia coli. Curr. Opin. Biotechnol. **19**(5), 414–419 (2008)
6. D.-K. Ro, E.M. Paradise, M. Ouellet, K.J. Fisher, K.L. Newman, J.M. Ndungu, K.A. Ho, R.A. Eachus, T.S. Ham, J. Kirby, M.C.Y. Chang, S.T. Withers, Y. Shiba, R. Sarpong, J.D. Keasling, Production of the antimalarial drug precursor artemisinic acid in engineered yeast. Nature **440**(7086), 940–943 (2006)
7. M.A. Marchisio, J. Stelling, Computational design tools for synthetic biology. Curr. Opin. Biotechnol. **20**(4), 479–485 (2009)
8. H. de Jong, Modeling and simulation of genetic regulatory systems: a literature review. J. Comput. Biol. **9**(1), 67–103 (2002)
9. M. Hucka, A. Finney, H.M. Sauro, H. Bolouri, J.C. Doyle, H. Kitano, A.P. Arkin, B.J. Bornstein, D. Bray, A. Cornish-Bowden, A.A. Cuellar, S. Dronov, E.D. Gilles, M. Ginkel, V. Gor, I.I. Goryanin, W.J. Hedley, T.C. Hodgman, J.H. Hofmeyr, P.J. Hunter, N.S. Juty, J.L. Kasberger, A. Kremling, U. Kummer, N. Le Novére, L.M. Loew, D. Lucio, P. Mendes, E. Minch, E.D. Mjolsness, Y. Nakayama, M.R. Nelson, P.F. Nielsen, T. Sakurada, J.C. Schaff, B.E. Shapiro, T.S. Shimizu, H.D. Spence, J. Stelling, K. Takahashi, M. Tomita, J. Wagner, J. Wang, The systems biology markup language (SBML): a medium for representation and exchange of biochemical network models. Bioinformatics **19**(4), 524–531 (2003)
10. B. Bartley, J. Beal, K. Clancy, G. Misirli, N. Roehner, E. Oberortner, M. Pocock, M. Bissell, C. Madsen, T. Nguyen, Z. Zhang, J.H. Gennari, C. Myers, A. Wipat, H. Sauro, Synthetic biology open language (SBOL) Version 2.0.0. J. Integr. Bioinform. **12**(2), 272 (2015)
11. H. McAdams, L. Shapiro, Circuit simulation of genetic networks. Science **269**(5224), 650–656 (1995)
12. D. Bernardi, J.T. Dejong, B.M. Montoya, B.C. Martinez, Bio-bricks: biologically cemented sandstone bricks. Constr. Build. Mater. **55**, 462–469 (2014)
13. *SBML Software Matrix* (2010). http://sbml.org/SBML_Software_Guide/SBML_Software_Matrix
14. A.A.K. Nielsen, B.S. Der, J. Shin, P. Vaidyanathan, V. Paralanov, E.A. Strychalski, D. Ross, D. Densmore, C.A. Voigt, Genetic circuit design automation. Science **352**(6281), aac7341–aac7341 (2016)
15. C.J. Myers, N. Barker, K. Jones, H. Kuwahara, C. Madsen, N.P.D. Nguyen, iBioSim: a tool for the analysis and design of genetic circuits. Bioinformatics **25**(21), 2848–2849 (2009)

Chapter 2
Fundamentals of Molecular Biology and Genetic Circuits

A biological system is composed of living organisms which consist of one or more living cells. The behavior of each of these cells is controlled by genetic circuits which perform dedicated tasks to achieve the overall functionality of a biological system. These genetic circuits, which are composed of several biological components (called the genes network), regulate the amount of proteins in a cell. This gene-regulated network is triggered by external signals, for example, light, temperature, presence of specific proteins, etc., to control the behavior of a living cell, effectively exhibiting a Boolean logic function. The aim of this chapter is to briefly introduce genetic circuits to the audience not familiar with synthetic biology. Section 2.1 gives a brief overview of biology and some basic terminologies, frequently used in genetic design, which are necessary to be known especially by the computer scientists or engineers who do not have primary knowledge of synthetic biology. Next, Sect. 2.2 presents an example of regulated transcription in lac operon and explains its genetic logic. Section 2.3 gives more information on the *standards* and Sect. 2.4 gives a brief overview of GDA tools.

2.1 Central Dogma of Molecular Biology

The *living cell* is the smallest biological unit of any living organism and is often called the *building block of life*. Each cell is composed of several organelles like mitochondria, ribosomes, nucleus, etc. The *nucleus* is the largest cellular component which contains part or all of the cell's genetic information. This genetic information is stored in the *deoxyribonucleic acid* (DNA) molecule, which is packaged into a thread-like structure called *chromosomes*. DNA is further divided into a group of nucleotide sequences called *genes*. Figure 2.1 shows the relationship between the eukaryotic cell's nucleus, chromosomes in the nucleus, and genes.

© Springer Nature Switzerland AG 2020
H. Baig, J. Madsen, *Genetic Design Automation*,
https://doi.org/10.1007/978-3-030-52355-8_2

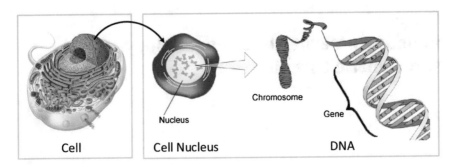

Cell   Cell Nucleus   DNA

Fig. 2.1 The hierarchical relationship of living cell and gene. (Image courtesy of BBC Science: http://www.bbc.co.uk/schools/gcsebitesize/science/edexcel/classification_inheritance/ genesandinheritancerev1.shtml)

DNA is composed of two nucleotide strands coiled around each other to form a double-helix structure. Each of these strands contain a sequence of four nucleobases: *cytosine (C), guanine (G), adenine (A)*, and *thymine (T)*. These bases on both of the strands bind to each other in pairs such that *A* only binds with *T* and *G* only binds with *C*. The sequence of these base-pairs codes for various genetic components, including promoters, operators, genes, etc.

Each gene is a region of DNA, which generates a specific protein through the processes called *transcription* followed by *translation*. During transcription, the particular region of DNA (gene) is copied into *ribonucleic acids* (RNAs) by another RNA molecule called *RNA polymerase* (RNAP), which binds to a specific region of that gene called the *promoter*. RNA polymerase then moves along the gene's coding sequence and temporarily breaks the bond between the two DNA strands causing them to unwind or unzip. During this unwinding process, the RNA transcript is generated as shown in Fig. 2.2. This process of RNAP generation continues until the moving RNAP reaches a region of DNA called *terminator*. At this instant, RNAP leaves DNA and the newly formed RNA is released. Many of these RNAs hold the instructions for constructing protein and hence are commonly termed as *messenger RNAs* (mRNAs).

Now, during the process of translation, another protein, *the ribosome*, binds to mRNA at its specific region, called the *ribosome-binding site*. The ribosome moves along mRNA and generates the specific proteins. This process of converting DNA into mRNA through transcription and then the conversion of mRNA into protein through translation is known as the *central dogma of molecular biology* [1].

There are two types of gene expressions: *constitutive* and *regulated*. A gene is expressed constantly in constitutive type of gene expression, whereas it is controlled and dependent on the environmental changes in regulated gene expression. The genetic circuits are based on the genes which are transcribed through regulation. This transcriptional regulation is carried out by regulatory proteins, called *transcriptional factors*, which bind to an *operator site*, a region of DNA near promoter. The transcription factor either blocks (referred to as *repressor*) or helps (referred to as *activator*) RNAP to bind to the promoter region to initiate the process of transcription.

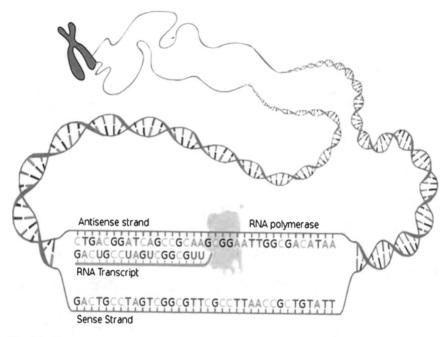

Fig. 2.2 The process of transcription. (Image courtesy of the National Human Genome Research Institute)

2.2   Example Genetic Circuit: Lac Operon

One of the classical systems used to investigate the transcriptional regulation of *lac operon* was presented by Jacob et al. in [2]. *Operon* is referred to as a region of DNA which consists of a group of genes controlled by a single promoter. Lac operon (or lactose operon) is required for the transport and metabolism of lactose in the bacterium *Escherichia coli*, and it was the first gene-regulatory network to be explored clearly.

Figure 2.3a shows the structure of the lac operon. Three genes, *lacZ, lacY,* and *lacA*, are required, as a cluster, to utilize lactose by the bacterium. The *lacZ, lacY,* and *lacA* genes code for the enzymes *beta-galactosidase, lactose permease,* and *galactoside transacetylase,* respectively. The *lacP* is the promoter region which transcribes the *lacZ, lacY,* and *lacA* genes as a single *polycistronic* mRNA. The *lacO* region is an operator site to which a transcription factor binds to regulate gene expression. The complete unit consisting of the lac promoter (*lacP*), lac operator (*lacO*), and the three genes (*lacZ, lacY,* and *lacA*) is known as the *lac operon*. The *lacI* is the regulatory gene of lac operon that codes for an mRNA that is translated to produce a protein known as *lac repressor*. The "T" (shown as red region) corresponds to the terminator region where the RNAP stops transcription. The black region between *lacI* gene and the lac promoter is the *activator-binding*

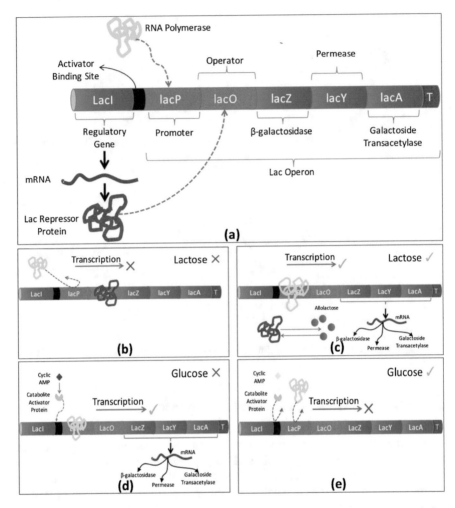

Fig. 2.3 Transcriptional regulation of lac operon. (**a**) Structure of lac operon. No transcription when (**b**) lactose is absent and (**e**) glucose is present. Transcription begins when (**c**) lactose is present and (**d**) glucose is absent

site (ABS), which helps RNA polymerase (shown as yellow structure) to bind to the promoter site.

As shown in Fig. 2.3b, when the lactose is not available inside the cell, the lac repressor recognizes the operator site and binds to it tightly. This prohibits the RNA polymerase to recognize the promoter region and thus prevents the operon to be transcribed. When lactose enters the cell, a small amount of it is converted to *allolactose*, which binds to the lac repressor. This causes a structural change in the lac repressor protein that prevents it from binding to the lac operator site. When lac repressor is not bound to an operator site, RNAP easily binds to a promoter region and transcribes the polycistronic mRNA. This mRNA is then translated to produce beta-galactosidase, lactose permease, and galactoside transacetylase proteins, as shown in Fig. 2.3c.

The discussion of Fig. 2.3b, c indicates that the lac operon is transcribed when lactose is present inside the cell. However, the binding of RNAP, to a promoter site, weakly depends on the presence of lactose, and strongly on the presence of the *catabolite activator protein* (CAP) inside the cell. CAP attaches to an *ABS* and helps RNAP to bind to the promoter region to drive high levels of transcription. The CAP cannot directly bind to an ABS, rather it is regulated by a small molecule known as *cyclic AMP* (cAMP), which acts as a "hunger signal" when the glucose levels inside the cell are low. Therefore, when the glucose level is low, the cAMP binds to CAP and enables it to attach to the ABS. When CAP attaches to the activation binding site, it helps RNAP to bind to the promoter region strongly and begin transcription. This process is shown in Fig. 2.3d.

On the contrary, when the glucose level rises, it reduces the concentration of cAMP, which in turn makes the CAP unable to attach to the ABS. Without CAP being attached to the ABS, RNAP cannot attach to the promoter region and thus the transcription process is stopped, as shown in Fig. 2.3e.

2.2.1 Genetic Logic in lac operon

From the discussion above, the natural genetic logic that exists in the transcriptional behavior of lac operon can be extracted. Figure 2.4a summarizes the logical behavior of lac operon in the form of truth table, with inputs being *Glucose* (G) and *Lactose* (L), and the output being *Transcription* (T) of *lacZ*, *lacY*, and *lacA* genes.

When glucose is absent (logic 0), CAP binds to ABS and RNAP should perform transcription. However, when the lactose is also absent (logic 0), the *lacI repressor* protein binds to the operator region and blocks RNAP to move along the DNA strand to perform transcription. Therefore, transcription is always blocked whenever

Inputs		Output
G	L	T
0	0	0
0	1	1
1	0	0
1	1	0

(a)

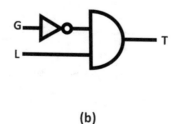

(b)

Fig. 2.4 Genetic logic in lac operon. (**a**) Truth table. (**b**) Circuit schematic

lactose is absent in the cell. Similarly, when glucose is present in the cell, it prohibits CAP to bind to ABS and thus reduces the affinity of RNAP to bind to the promoter region to begin transcription. However, when glucose is absent and lactose is present, the lac operon is transcribed, resulting in the Boolean logic shown as a circuit diagram in Fig. 2.4b.

2.2.2 The Standard SBOL Representation of lac operon

Figure 2.4b gives a fairly low-level detail of how lac operon works, which is not the standard way of representing any genetic system. As mentioned earlier in Chap. 1, *SBOL* is the standard way to represent the high-level diagrams of genetic systems. The equivalent SBOL visual (or SBOLv) diagram of the genetic system for lac operon would be something similar to the diagram shown in Fig. 2.5. This figure indicates that a single promoter, P_{lac}, is responsible for the transcription of three genes, *lacZ, lacY*, and *lacA*. The presence of glucose represses and the presence of lactose activates the promoter, reflecting that a transcription is initiated when glucose is absent and lactose is present in the cell. The symbol "T" represents the terminator region of DNA, where the transcription is stopped. As a result of transcription, an mRNA is produced and the *ribosomes* bind to this mRNA at the *ribosome-binding site* (shown as semisphere in Fig. 2.5), to carry out the production of output proteins *beta-galactosidase, permease*, and *galactoside transacetylase*, from the genes *lacZ, lacY*, and *lacA*, respectively.

The example discussed above is a natural genetic logic circuit which exists in the lac operon. Some researchers have already started to engineer custom genetic logic components to achieve the desired logical behavior in a living cell [3, 4]. We have used the SBML models of these genetic logic circuits [3, 4] to test the tools and method presented in this book.

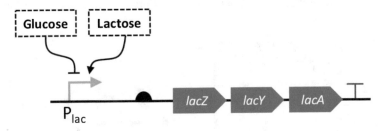

Fig. 2.5 SBOL visual representation (or SBOLv) of lac operon genetic system

2.3 Standards

It is mentioned in Chap. 1 that SBML and SBOL are two major standards to represent genetic model's behavior and structure, respectively. Both of these standards are developed by the members of systems and synthetic biology communities.

 SBOL is used to describe the structure of a model, whereas SBML is used to describe the behavior.

The purpose of SBML standard is to exchange essential aspects of biological model among different software and is supported by over 290 GDA tools. It is a machine-readable *eXtensible Markup Language* (XML) that is independent of any specific software language. For example, the SBML model of a genetic AND gate circuit, designed in iBioSim [5], is shown in Fig. 2.6b. Tools which support SBML synthesis allow users to define the model parameters, species, and their biochemical interactions using mathematical expressions. As an example, the SBML model of a genetic AND gate circuit is shown in Fig. 2.6b in which the reaction for $pTac$, along with an external influence of IPTG inducer, to produce $A1AmtR\_protein$ is shown in Fig. 2.6c. The values of the parameters in this figure, for example, kb, ko_r, nc, etc., are defined separately. Different SBML-synthesis tools may have their own icons to represent standard biological processes like *repression, activation*, etc. For example, in Fig. 2.6b, $A1AmtR\_protein$ repressing the following promoter $pAmtR$ is shown with the red line having round-headed circle. The same processes are represented differently in different tools.

In order to keep the uniformity in representing these models, the SBOL is developed to document all biological models in a standardized manner. It is an emerging data standard for synthetic biology with growing support among several GDA software tools, including biochemical modeling tools [6–8], design composition tools [6, 7, 9–11], and sequence editing tools [12, 13]. It is also an *eXtensible* standard, so it can easily adapt the evolving needs of the synthetic biology community. Figure 2.6d shows the SBOLv diagram of the genetic AND gate model shown in Fig. 2.6b.

As said before different tools may have their own icons to represent the biological processes and species, but all of them are supposed to generate the same *XML* document in order to be used by other software tools. Figure 2.7a, b shows the cropped images of SBML and SBOL *xml* files, respectively, of the same genetic AND gate model shown in Fig. 2.6. Both of the images shown in this figure depict some portion of these SBML and SBOL xml files, showing how the reaction (between $pTac$ and $A1AmtR\_protein$) shown in Fig. 2.6c is represented.

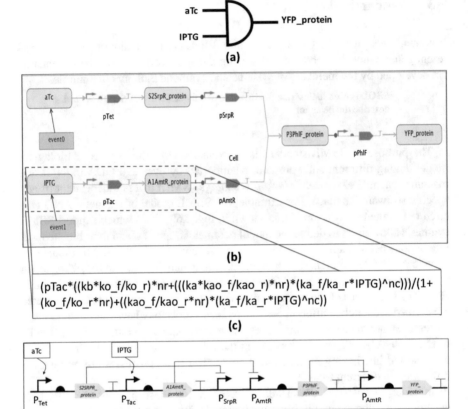

Fig. 2.6 SBML and SBOLv diagram of genetic AND gate. (**a**) Circuit schematic. (**b**) SBML model design in iBioSim [5]. (**c**) Example kinetic reaction. (**d**) SBOLv diagram

2.4 Genetic Design Automation (GDA) Tools

Numerous computational tools [7, 14–16] have been developed to assist users in designing genetic circuits through an automated process. These tools includes DNA sequence editing, biochemical modeling, design composition, and technology mapping tools. According to [16], there are more than 290 tools which support SBML model construction and simulation; and about 30 of them are GDA tools which support sequence editing, design composition, optimization, and technology mapping. Some of these tools serve as a toolbox for commercial platforms, including MATLAB, Mathematica, and Oracle; some are developed as APIs or plug-ins to specific software systems, while others are independent tools for design and simulation.

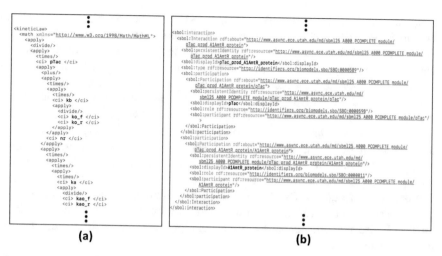

Fig. 2.7 Cropped images of SBML and SBOL files of genetic AND gate circuit shown in Fig. 2.6.
(**a**) SBML file. (**b**) SBOL file

2.4.1   Sequence Editing Tools

Sequence editing tools are typically considered low-level tools which enable user to
construct, edit, or annotate the base-pair sequences of genetic components. These
tools include SBOLDesigner [17], Synthetic GeneDesigner [12], GeneDesign [18],
VectorEditor [13], and Kera [19]. Only few of the sequence editing tools support the
SBOL standard. However, nearly all of these tools support reading and writing plain
text DNA sequences that comply with the *International Union of Pure and Applied
Chemistry* (IUPAC) codes for *nucleotides* [20] and *amino acids*.

2.4.2   Biochemical Modeling and Design Composition Tools

These tools allow users to develop mathematical models and usually also provide
users with the ability to analyze these models. They require users to manually
compose designs and generate standard files automatically. Some of these tools are
CellDesigner [21], BioUML [22], iBioSim [5], COPASI [23], D-VASim [24, 25],
Uppaal [26], Asmparts [27], GEC [28], GenoCAD [10], Kera [19], ProMoT [29],
Antimony [30], Proto [7], SynBioSS [31], TinkerCell [6], and Cello [3]. Few of
these tools also support model checking of genetic circuits, including [5, 26, 32–34].

A vast majority of these tools provides different level of support of reading and/or
writing SBML files to capture the mathematical behavior of biological models. For
example, CellDesigner, iBioSim, COPASI, and ProMoT support both import and
export of SBML files; Asmparts, GenoCAD, GEC, and SynBioSS, on the other

hand, are only able to export SBML; while D-VASim, being a simulation tool, is able to import SBML files only. Each of these tools have their own advantages. For example, iBioSim, Antimony, and BioUML tools support hierarchical model composition in a standardized SBML format. iBioSim also supports in designing and modifying the DNA sequences of genetic constructs using SBOL data model [17]. Most of these tools support both deterministic and stochastic simulations, including iBioSim, COPASI, and D-VASim. A noticeable feature of SynBioSS is that it supports running the complex models on a *supercomputer* to speed up the process of finding genetic parts to construct a desired biological system. An apparent feature of TinkerCell tool is that it provides a platform to integrate third-party algorithms for testing different methods relevant to synthetic biology. The distinctive feature of D-VASim is its virtual simulation environment that lets the users perform experimentation by dynamically interacting with the biological models. It also supports *automated* timing and logic analysis of genetic circuits in its virtual simulation platform.

Some of the modeling tools support programming languages and allow the user to design biological circuits using different programming formats. For example, *Proto* converts high-level program specifications into gene-regulatory networks, optimizes it, and then validates their behavior through simulation. *Kera* is a C-like object-oriented programming language, which supports a biopart rule library called *Samhita* which is a database containing part IDs, their types, and sequences. *GenoCAD*, a web-based tool, also supports some features of programming language and a built-in database of BioBricks. It allows the user to perform model simulation using *mass action kinetics*. *Antimony* is a module-based programming language, which provides a special syntax to create modular genetic networks. The additional library, *libAntimony*, in this tool allows other software packages to import models and convert them to SBML. Unlike GenoCAD, which uses rates of mass action for simulation, Antimony uses the rates of gene expression, for example, *polymerases per second* (PoPS) and *ribosomes per second* (RiPS) for simulations. Since it is difficult to connect genetic gates because of different input and output proteins, the *rate of gene expression* is often used as the integration signal to solve this problem [4]. The rate of gene expression, which is similar to the rate of flow of electron (electric current), is the rate at which the RNA polymerases move across a DNA strand.

GEC, developed by Microsoft Corporation, is a rule-based programming language, which also allows the user to express the logical interaction of biological components in a modular manner. The GEC program is translated into multiple possible genetic devices (RBS, promoters, etc.). The GEC model is refined by simulating these possible devices iteratively and ruling out those with undesired simulation results. A recently published web-based tool, named *Cello*, allows the user to develop genetic circuits using *verilog* (http://www.verilog.com/)—a hardware description language used for designing electronic circuits. The circuits are developed using the genetic gates which have already been tested in the laboratory.

2.4.3 Genetic Mapping Tools

Genetic technology mapping tools automatically select genetic components from library and integrate them together to achieve the desired functionality of a genetic circuit. The mapping techniques, similar to those adapted in EDA and software engineering, are mostly used to perform technology mapping of genetic circuits. However, technology mapping of genetic circuit is much more computationally intensive as compared to the technology mapping of electronic circuits. It is because the genetic components have different input and output signals as opposed to electronic components which have the same physical quantity both at input and output, i.e., the voltage. Due to this, it is challenging to search and connect the right components together which should not only be compatible with each other but also avoid unwanted *crosstalk* to achieve the desired functionality.

Some of the tools which support genetic technology mapping are BioJADE [35], GEC [28], MatchMaker [36], SBROME [37], iBioSim [38], GeneTech [39], etc. BioJADE and GEC use *exact methods* to find the optimal solutions. On other hand, both MatchMaker and SBROME use *heuristic methods* to find all possible solutions quickly and then rank them by quality. iBioSim uses a similar *Directed Acyclic Graph* (DAG)-based approach for technology mapping, which is used in EDA [40]. It first generates the DAG and then performs matching and covering to obtain the optimal solution. This tool is based on two assumptions: first, the circuits generated from the library do not have feedback, and second, the circuit components can be connected together if their molecular signals at input and output are the same. However, GeneTech generates all possible solutions using *depth-first search* approach. It starts off by mapping the components at input stage first and then searches for the right components (from the gates library) that can be connected in succession as a second stage, and so on. Unlike the first assumption of iBioSim, GeneTech avoid using such components which create unintended feedback loops.

References

1. F. Crick, Central dogma of molecular biology. Nature **227**(5258), 561–563 (1970)
2. F. Jacob, J. Monod, Genetic regulatory mechanisms in the synthesis of proteins. J. Mol. Biol. **3**, 318–56 (1961)
3. A.A.K. Nielsen, B.S. Der, J. Shin, P. Vaidyanathan, V. Paralanov, E.A. Strychalski, D. Ross, D. Densmore, C.A. Voigt, Genetic circuit design automation. Science **352**(6281), aac7341–aac7341 (2016)
4. C.J. Myers, *Engineering Genetic Circuits* (Chapman and Hall/CRC, New York, 2009)
5. C.J. Myers, N. Barker, K. Jones, H. Kuwahara, C. Madsen, N.P.D. Nguyen, iBioSim: a tool for the analysis and design of genetic circuits. Bioinformatics **25**(21), 2848–2849 (2009)
6. D. Chandran, F.T. Bergmann, H.M. Sauro, TinkerCell: modular CAD tool for synthetic biology. J. Biol. Eng. **3**(1), 19 (2009)
7. J. Beal, T. Lu, R. Weiss, Automatic compilation from high-level biologically-oriented programming language to genetic regulatory networks. PLoS One **6**(8), e22490 (2011)

8. C. Madsen, C.J. Myers, T. Patterson, N. Roehner, J.T. Stevens, C. Winstead, Design and test of genetic circuits using iBioSim. IEEE Des. Test Comput. **29**(3), 32–39 (2012)

9. J. Chen, D. Densmore, T.S. Ham, J.D. Keasling, N.J. Hillson, DeviceEditor visual biological CAD canvas. J. Biol. Eng. **6**(1), 1 (2012)

10. Y. Cai, M.L. Wilson, J. Peccoud, GenoCAD for IGEM: a grammatical approach to the design of standard-compliant constructs. Nucleic Acids Res. **38**(8), 2637–2644 (2010)

11. G. Misirli, J.S. Hallinan, T. Yu, J.R. Lawson, S.M. Wimalaratne, M.T. Cooling, A. Wipat, Model annotation for synthetic biology: automating model to nucleotide sequence conversion. Bioinformatics **27**(7), 973–979 (2011)

12. A. Villalobos, J.E. Ness, C. Gustafsson, J. Minshull, S. Govindarajan, Gene Designer: a synthetic biology tool for constructing artificial DNA segments. BMC Bioinf. **7**, 285 (2006)

13. T.S. Ham, Z. Dmytriv, H. Plahar, J. Chen, N.J. Hillson, J.D. Keasling, Design, implementation and practice of JBEI-ICE: an open source biological part registry platform and tools. Nucleic Acids Res. **40**(18), e141–e141 (2012)

14. J.T. MacDonald, C. Barnes, R.I. Kitney, P.S. Freemont, G.-B.V. Stan, Computational design approaches and tools for synthetic biology. Integr. Biol. Quan. Biosci. Nano Macro **3**(2), 97–108 (2011)

15. G. Setti, M. di Bernardo, H. Koeppl, D. Densmore, in *Design and Analysis of Biomolecular Circuits: Engineering Approaches to Systems and Synthetic Biology* (Springer, Berlin, 2011)

16. *SBML Software Matrix* (2010). http://sbml.org/SBML_Software_Guide/SBML_Software_Matrix

17. M. Zhang, J.A. McLaughlin, A. Wipat, C.J. Myers, SBOLDesigner 2: an intuitive tool for structural genetic design. ACS Synth. Biol. **6**(7), 1150–1160 (2017)

18. S.M. Richardson, S.J. Wheelan, R.M. Yarrington, J.D. Boeke, GeneDesign: rapid, automated design of multikilobase synthetic genes. Genome Res. **16**(4), 550–556 (2006)

19. P. Umesh, F. Naveen, C.U.M. Rao, A.S. Nair, Programming languages for synthetic biology. Syst. Synth. Biol. **4**(4), 265–269 (2010)

20. H.B.F. Dixon, H. Bielka, C.R. Cantor, Nomenclature for incompletely specified bases in nucleic acid sequences: Recommendations 1984. Nucleic Acids Res. 13(9), 3021 (1986)

21. A. Funahashi, Y. Matsuoka, A. Jouraku, M. Morohashi, N. Kikuchi, H. Kitano, CellDesigner 3.5: a versatile modeling tool for biochemical networks. Proc. IEEE **96**(8), 1254–1265 (2008)

22. F. Kolpakov, M. Puzanov, A. Koshukov, *BioUML: Visual Modeling, Automated Code Generation and Simulation of Biological Systems*, vol. 3 (2006)

23. S. Hoops, R. Gauges, C. Lee, J. Pahle, N. Simus, M. Singhal, L. Xu, P. Mendes, U. Kummer, COPASI: a complex pathway simulator. Bioinformatics **22**(24), 3067–3074 (2006)

24. H. Baig, J. Madsen, D-VASim: an interactive virtual laboratory environment for the simulation and analysis of genetic circuits. Bioinformatics **33**(2), 297–299 (2017)

25. H. Baig, J. Madsen, Simulation approach for timing analysis of genetic logic circuits. ACS Synth. Biol. **2**, acssynbio.6b00296 (2017)

26. J. Bengtsson, K. Larsen, F. Larsson, P. Pettersson, W. Yi, UPPAAL-a tool suite for automatic verification of real-time systems, in *Lecture Notes in Computer Science (Including Subseries Lecture Notes in Artificial Intelligence and Lecture Notes in Bioinformatics)*, vol. 1066 (1996), pp. 232–243

27. G. Rodrigo, J. Carrera, A. Jaramillo, Asmparts: assembly of biological model parts. Syst. Synth. Biol. **1**(4), 167–170 (2007)

28. M. Pedersen, A. Phillips, Towards programming languages for genetic engineering of living cells. J. R. Soc. Interface **6**(Suppl 4), 437–50 (2009)

29. S. Mirschel, K. Steinmetz, M. Rempel, M. Ginkel, E.D. Gilles, ProMoT: modular modeling for systems biology. Bioinformatics **25**(5), 687–689 (2009)

30. L.P. Smith, F.T. Bergmann, D. Chandran, H.M. Sauro, Antimony: a modular model definition language. Bioinformatics **25**(18), 2452–2454 (2009)

31. A.D. Hill, J.R. Tomshine, E.M.B. Weeding, V. Sotiropoulos, Y.N. Kaznessis, SynBioSS: the synthetic biology modeling suite. Bioinformatics **24**(21), 2551–2553 (2008)

32. K.G. Larsen, A. Legay, M. Mikucionis, P. Bulychev, A. David, D.B. Poulsen, Checking and distributing statistical model checking, in *Proceedings of the 4th NASA Formal Methods Symposium (LNCS 7226)* (Springer, Berlin, 2012), pp. 449–463
33. S. Jha, E. Clarke, C. Langmead, A. Legay, A. Platzer, P. Zuliani, A Bayesian approach to model checking biological systems. Lect. Notes Comput. Sci. **5688**(2005), 218–234 (2009)
34. E.M. Clarke, J.R. Faeder, C.J. Langmead, L.A. Harris, S.K. Jha, A. Legay, Statistical Model Checking in BioLab: applications to the automated analysis of T-cell receptor signaling pathway, in *Proceedings of the Computational Methods in Systems Biology: 6th International Conference CMSB 2008, Rostock, Germany, October 12–15, 2008*. LNCS (Springer, Berlin, 2008), pp. 231–250
35. J.a. Goler, BioJADE: a design and simulation tool for synthetic biological systems. DSpace@MIT is a digital repository for MIT's research, including peer-reviewed articles, technical reports, working papers, theses, and more, 54 (2004). http://hdl.handle.net/1721.1/30475
36. F. Yaman, S. Bhatia, A. Adler, D. Densmore, J. Beal, Automated selection of synthetic biology parts for genetic regulatory networks. ACS Synth. Biol. **1**(8), 332–344 (2012)
37. L. Huynh, A. Tsoukalas, M. Koppe, I. Tagkopoulos, SBROME: a scalable optimization and module matching framework for automated biosystems design. ACS Synth. Biol. **2**(5), 263–273 (2013)
38. N. Roehner, C.J. Myers, Directed acyclic graph-based technology mapping of genetic circuit models. ACS Synth. Biol. **3**(8), 543–555 (2014)
39. H. Baig, J. Madsen, A top-down approach to genetic circuit synthesis and optimized technology mapping, in *Proceedings of the 9th IWBDA* (2017), pp. 28–29
40. K. Keutzer, DAGON: technology binding and local optimization by DAG matching, in *Proceedings of the 24th ACM/IEEE Design Automation Conference* (1987), pp. 341–347

Part II
Virtual Experimentation and Technology Mapping of Genetic Circuits

In this part, you will be introduced to methods and tools for the virtual experimentation, analysis, and synthesis of genetic logic circuits. This part consists of four chapters.

Chapter 3
Virtual Experimentation Using *DVASim*

Simulation and behavioral analysis of genetic circuits is a standard approach of functional verification prior to their physical implementation. Many software tools have been developed to perform in-silico analysis for this purpose, but none of them allows users to interact with the model during run-time. The run-time interaction gives the user a feeling of being in the lab performing a real-world experiment. In this chapter, you will be introduced with a tool named DVASim (Dynamic Virtual Analyzer and Simulator) which allows users to perform *virtual laboratory* experimentation. This tool provides a user-friendly environment to simulate and analyze the behavior of genetic logic circuit models represented in an SBML format. Hence, SBML models developed in other software environments can be analyzed and simulated in DVASim.

3.1 Motivation

In the wet lab, biologists are either provided with the ready-made biological model available in a test tube or are given a specification/recipe from which to prepare the model in the lab. Their duty is to analyze the model and verify its functional behavior. This analysis is done interactively, among other things, by increasing the molecular concentration of input species at any instant of time and observing the effects. However, this is a very time consuming task as the time taken by lab experiments to complete varies between few hours to couple of days. Furthermore, the experiments are often required to be repeated again in case of mishandling the apparatus or when the parameters are not selected appropriately.

Electronic Supplementary Material The online version of this chapter (https://doi.org/10.1007/978-3-030-52355-8_3) contains supplementary material, which is available to authorized users.

This motivated the developers to come up with an idea of developing *in-silico* virtual laboratory environment where a user can perform interactive experiments by varying the molar concentration of external signals and observe the effects during run-time. Hence the tool, DVASim (Dynamic Virtual Analyzer and Simulator), came into being. DVASim offers deterministic as well as stochastic simulation and differs from other software tools by being able to extract and validate the Boolean logic from the SBML model. DVASim is also capable of analyzing the threshold value and propagation delay of a genetic circuit model.

The SBML and Cell Mark-up Language (CellML) [1] are the two standard methods of representing biological models in a machine-readable form, which enable models to be shared and published in a way that can be used by different software tools. SBML is supported by most relevant tools for synthetic biology. An SBML file holds the model information including species, reaction parameters, kinetic laws, initial concentrations, etc. Beside these modular descriptions of a bio model, SBML also allows a user to model a sequence of input patterns in order to capture more behavioral details. This is done through *events*, which describe the instantaneous, discontinuous state changes in the model [2]. For example, in genetic circuits, events are used to trigger the concentration of any input species to a certain level, at a specific point in time, and to observe the effects on the concentration of output species. Since events are predefined and cannot be changed during run-time, the output of a genetic logic circuit will be analyzed only for defined events. In order to observe the complete behavior of an output of a genetic logic circuit, a different set of input conditions, i.e., when to change what input to which level, must be defined in each event. Even for moderate sized genetic logic circuits, capturing all the combinations of inputs and concentration levels may require a very large number of events to be defined and simulated.

On the other hand, a run-time interaction capability with the model is more suitable to make direct changes in the concentration of input species at any instant of time to observe the model's behavior. This not only helps the user to analyze the model easily by triggering the concentration of input species to any level and at any instant of time, but also makes a user free of defining a long list of events for all the possible combinations of inputs in the SBML description. Furthermore, the interactive simulation enables a user to receive feedback in parallel with their experimental intervention, which enables certain types of learning and optimization that would not be possible otherwise.

Besides giving a biologist the feeling of being in the lab, DVASim has also been proven to be useful in helping early-stage researchers, students, and other users with less experience of biology, to get an intuitive feeling of the underlying biological processes and their interactions. DVASim plays a vital role for educating inexperienced users to observe the live biological phenomenon in a virtual laboratory environment without being afraid of overreaction and mishandling of the apparatus.

3.2 Experimental Approach

DVASim is developed on the LabVIEW (Laboratory Virtual Instrument Engineering Workbench) programming platform, which is a graphical programming language (discussed in Chap. 7) commonly used for rapid development of instrumentation systems for data acquisition, instrument control, and industrial automation (www.ni. com). The basic flow of the virtual simulation and analysis environment in DVASim is shown in Fig. 3.1.

Figure 3.1 shows that the SBML model of a genetic circuit is first loaded in DVASim, and the components of an SBML model can be optionally analyzed in a user-friendly manner. Then a user can generate a separate virtual instrument (VI) to perform ODE and stochastic simulations. These VIs help a user to observe

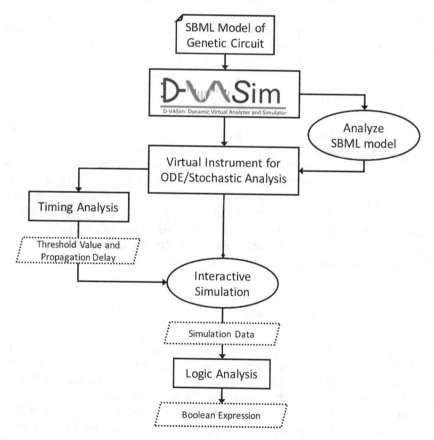

Fig. 3.1 A work flow diagram of DVASim showing how it can be used for simulation, analysis, and verification of genetic logic circuit models. Elliptical nodes represent the steps to be performed by a user, rectangular nodes represent the automated processes, and the dotted-parallelogram shows the output from the previous stage

the reactions graphically and interact with the model at run-time. This process is equivalent to setting up an apparatus for testing and experimentation of a model in the wet lab.

For stochastic simulations, the timing analysis of a genetic circuit can also be performed if they are not known. The timing analysis helps a user to analyze the threshold value and the propagation delay of a genetic circuit model. If a user already knows the threshold value and propagation delay of a circuit, the genetic circuit model can be analyzed by interacting with the model during run-time. This run-time interaction allows a user to change the concentration of input species and the parameter values at any instant of time and observe the change in circuit's behavior alongside.

Once all the possible input combinations are applied carefully, a user can analyze/verify the logical behavior of either a complete circuit, or any intermediate components of a circuit. The tool analyzes the simulation data and produces the result in the form of a Boolean expression.

The virtual instrument generated by DVASim, for stochastic and ODE both, also logs the simulation data for analysis and retrieval of the user-session at a later stage.

3.2.1   SBML Support

DVASim supports the SBML format level 3, version 1 (l3v1) core[1]. It processes the SBML file using the jSBML library [3], and extracts and presents the information of all components in a tabular format, as shown in Fig. 3.2. This figure shows the first interface which a user sees after initializing and loading the SBML model in DVASim. Each *tab* describes the corresponding SBML component in a user-friendly manner. For example, the *selected tab* shown in Fig. 3.2 depicts the information of *Reactions* defined in the SBML file of the genetic AND gate model [4]. It not only shows the kinetic reactions, but also the ordinary differential equations, generated by DVASim, to simulate the deterministic behavior of this model.

3.2.2   Virtual Instrumentation

Depending on the option selected, DVASim generates a virtual instrument for the deterministic or stochastic analysis separately. Once the instrument is generated, the user can analyze the model by varying those species concentrations, which act as external modifiers or external inputs. For example, selecting the **Generate SSA VI option** (shown in Fig. 3.2) generates the virtual instrument, shown in Fig. 3.3, to perform the stochastic analysis of the genetic logic circuit described in the SBML file.

---

[1]DVASim does not currently support SBML *packages*. The complete list of supported SBML components is declared in the ***Read Me*** file in the DVASim download package.

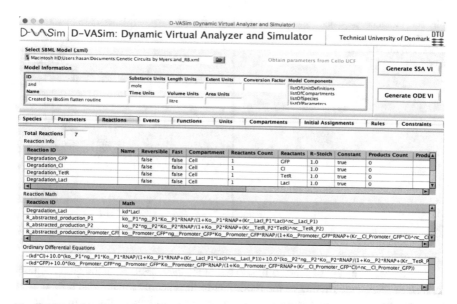

Fig. 3.2 The main interface of DVASim showing how the components of SBML file can be analyzed through a user-friendly interface

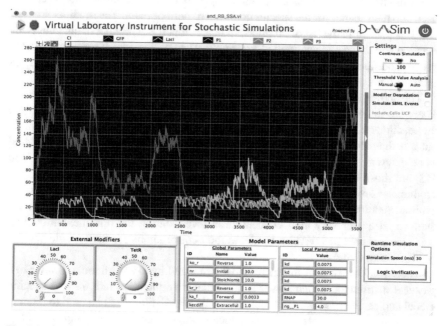

Fig. 3.3 Virtual instrument showing the stochastic simulation traces of the genetic AND gate (shown in Fig. 3.5) obtained from [4]

Fig. 3.4 Example reactions explaining the identification process of the external modifiers

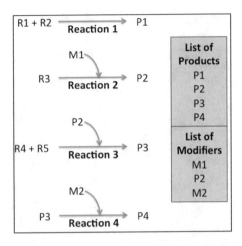

The SBML file contains the complete list of species involved in the circuit model as well as the species acting as the modifiers to the specific kinetic reaction. It does not, however, specify explicitly which species acts as external inputs to the entire circuit model. DVASim identifies the external species first by arranging the names of all modifiers species and the products species of each reaction in two separate lists, and then search for the species common to both of them. The modifier species, which is present in the list of products, indicates that it is a product of an underlying reaction and, hence, is not applied externally. The species, which are present in the list of modifiers but are not in the list of any reaction products, are those acting as external inputs (or external modifiers) to the circuit.

The above process is explained in the example list of reactions shown in Fig. 3.4. There are three species in this figure—M1, P2, and M2, which act as modifiers to the reactions 2, 3, and 4, respectively. The species P2 is modifying the reaction 3, but it is the product of the underlying reaction 2; therefore, it cannot be considered as an external modifier. The rest of the species in the list of modifiers i.e. M1 and M2, are those that are acting as external modifiers because none of them are the products of any of the reactions, and thus considered as external inputs. When the external modifiers or inputs are identified, DVASim creates the control knobs for them to let a user vary their concentration levels during the run-time simulation. It is also possible in DVASim to create the control knobs for the specific species (see DAVSim QSG).

The virtual instrument for each model generated by DVASim looks similar to a physical instrument, which serves as a standalone simulation tool for that specific model and can be used later without having its SBML file. It can be used to interact with the model by tuning the input concentration levels with the help of control knobs and observing the effects graphically. This run-time interaction with the model also helps users to trigger the concentration of input species to any level at any instant of time without defining the long list of events in an SBML file. In a similar way that input concentrations can be changed using control knobs, the parameters editor can be used to vary parameters, like degradation rate, temperature, etc., during run-time and observe their effects graphically. Furthermore, DVASim also allows a user to simulate events defined in the SBML file.

Unlike wet lab experimentation, a user may speed-up or slow-down the reactions (for stochastic simulation) with the help of numeric speed control displayed on each virtual instrument (see Fig. 3.3). Moreover, when the simulation is stopped by a user, the simulation data and the screen shot of a graphical window are stored automatically in the default application folder.

3.2.3 Virtual Experimentation

The Gillespie's direct stochastic simulation algorithm (SSA) [5] is implemented to capture the stochastic nature of the biological models described in the SBML file. For deterministic simulations[2], DVASim supports ten different types of continuous solvers (see DAVSim QSG (Quick Start Guide)).

Figure 3.5 shows the genetic AND gate circuit, constructed with the genetic NAND and NOT gates [4]. Figure 3.5a depicts the SBOLv diagram of genetic AND gate in which P_1, P_2, and P_3 corresponds to the promoter regions of DNA.

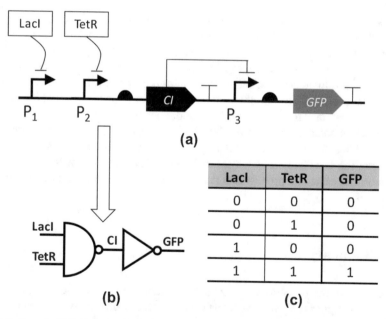

Fig. 3.5 Genetic AND gate circuit [4]. (**a**) SBOLv diagram. (**b**) Circuit schematic. (**c**) Truth table. Reprinted (adapted) with permission from [6]. Copyright (2017) American Chemical Society

---

[2]Deterministic simulation is tested with the first 400 cases of SBML benchmark suite, and it produced the correct results for all of the supported components in DVASim.

Figure 3.5b, c shows the circuit schematic and the truth table of genetic AND gate, respectively. In this example, when two proteins, *LacI* and *TetR*, are present in the significant amount within the cell, they inhibit promoters P_1 and P_2 to produce the output protein CI. When the concentration of CI falls below a certain level, promoter P_3 is activated and produces an output, i.e., green fluorescent protein (GFP).

The interactive stochastic simulation traces of the genetic AND gate model, in DVASim, is shown in Fig. 3.3. DVASim identified the external modifiers (or external inputs), *LacI* and *TetR*, and created control knobs to let a user control their concentration levels during run-time simulation. The simulation traces shown in Fig. 3.3 indicates that the user varied the concentration levels of input species at different instants of time and in different combinations. The concentration of *CI* is suppressed, when the concentrations of both LacI and TetR are present in a significant amount in the cell at the same time (between 2500–4700 time units). When the concentration of *CI* falls below a certain level, the promoter P_3 is activated and the GFP is produced, thus exhibiting the AND logic in the cell, as shown in Fig. 3.3.

3.2.4   Logic Verification and Timing Analysis

Besides interactive simulation of any SBML-based model, DVASim is also capable of verifying the logical behavior of a genetic circuit model by extracting the observed Boolean logic function from the simulation data. This functionality is useful in two ways—first, it allows a user to verify circuits, built by cascading several genetic logic circuits; secondly, it helps a user to extract the Boolean logic of a model even when a user does not have any prior knowledge about the model's expected behavior. The details of this functionality are presented in Chap. 5.

In order to obtain the correct Boolean logic, all possible input combinations must be applied in a significant amount to trigger the circuit's output. To determine which concentration level should be considered logic 0 and 1, DVASim is able to obtain the threshold value and propagation delay of a circuit, which is described in details in Chap. 4.

3.3   Summary

DVASim is in the process of continuous development and has been continuously upgraded to higher versions. The latest version of DVASim can be downloaded from http://bda.compute.dtu.dk/downloads/d-vasim/. The video demo of the complete simulation flow can be seen at http://bda.compute.dtu.dk/user-manuals/d-vasim/.

The next Chap. 4 describes how the threshold value and the propagation delay of these genetic circuit models can be analyzed. Such analyses can then be used to obtain the Boolean behavior (Chap. 5) of a genetic circuit model.

Problems

You are given five models (~/Resources/Problems/Chapter3/Models/) of different genetic circuits. Load each of these models one by one and attempt the following problems:

3.1 List down the names and number of species in each model.

3.2 How many reaction are there? What are the stoichiometries of reactants (R-Stoich) and products (P-Stoich) in each reaction. What do the reactant and product stoichiometries indicate in each reaction?

3.3 How many events are there? What are the values of *HighTime*, *HighTime2*, and *High*? What do these values indicate?

3.4 (Deterministic Simulation)

Click on **Generate ODE VI**. Go to **Simulation Settings** » Select **Runge–Kutta 45 (variable)** as *Continuous Solver* » Set **final time** to 6000 » Click **OK**. Now run the environment (by clicking on the green play button at the top-left corner) and observe the simulation results. Verify that the species are triggered to the right concentration level at the time specified by predefined parameters. Take the screenshot of this deterministic simulation.

 Screenshot in *ODE VI* is captured automatically when the stop button (red octagonal button on top-left) is pressed.

 Make sure all the relevant plots are visible by clicking on the plot legend curve and select **Plot Visible**. Also, differentiate the curves using different colors.

3.5 (Interactive Stochastic Simulation)

Go back to the main DVASim interface and now click on **Generate SSA VI**. Run the environment (by clicking on the green play button at the top-left corner). You may observe the random stochastic curves plotting on your screen which shows the current behavior of a model. Now interact with a model by increasing the input concentrations using control knobs. Observe the behavior by applying inputs in different combinations. Take the screenshot of your interaction with the model until 6000 time units.

 Screenshot in *SSA VI* is captured automatically when the stop button (red octagonal button on top-left) is pressed.

3.6 Turn on the **Modifier Degradation** option (on right-hand side) and run the virtual environment again. What difference do you observe now as compared to your interaction with the model in Problem 3.5?

3.7 How many files are generated under ~/Sim_Results/ directory related to each model? Apart from *.txt* file, what are the purposes of other generated files?

References

1. C.M. Lloyd, M.D.B. Halstead, P.F. Nielsen, CellML: its future, present and past, in *Progress in Biophysics and Molecular Biology*, vol. 85(2–3) (2004), pp. 433–450
2. M. Hucka, A. Finney, H.M. Sauro, H. Bolouri, J.C. Doyle, H. Kitano, A.P. Arkin, B.J. Bornstein, D. Bray, A. Cornish-Bowden, A.A. Cuellar, S. Dronov, E.D. Gilles, M. Ginkel, V. Gor, I.I. Goryanin, W.J. Hedley, T.C. Hodgman, J.H. Hofmeyr, P.J. Hunter, N.S. Juty, J.L. Kasberger, A. Kremling, U. Kummer, N. Le Novére, L.M. Loew, D. Lucio, P. Mendes, E. Minch, E.D. Mjolsness, Y. Nakayama, M.R. Nelson, P.F. Nielsen, T. Sakurada, J.C. Schaff, B.E. Shapiro, T.S. Shimizu, H.D. Spence, J. Stelling, K. Takahashi, M. Tomita, J. Wagner, J. Wang, The systems biology markup language (SBML): a medium for representation and exchange of biochemical network models. Bioinformatics **19**(4), 524–531 (2003)
3. A. Drager, N. Rodriguez, M. Dumousseau, A. Dorr, C. Wrzodek, N. Le Novere, A. Zell, M. Hucka, JSBML: a flexible java library for working with SBML. Bioinformatics **27**(15), 2167–2168 (2011)
4. C.J. Myers, *Engineering Genetic Circuits* (Chapman and Hall/CRC, New York, 2009)
5. D.T. Gillespie, A general method for numerically simulating the stochastic time evolution of coupled chemical reactions. J. Comput. Phys. **22**(4), 403–434 (1976)
6. H. Baig, J. Madsen, Simulation approach for timing analysis of genetic logic circuits. ACS Synth. Biol. **2**, acssynbio.6b00296 (2017)

Chapter 4
Genetic Circuits Timing Analysis

Analogous to microelectronics, where timing analysis is a crucial requirement for ensuring the correct operation of a logic circuit, the timing analysis of genetic logic circuits may become an essential design characteristic as well. The transistors, used in the composition of digital logic gates, have well-defined threshold voltage values [1], which categorize the logic levels 0 and 1. Hence, the timing characteristics, like propagation delay, hold time, setup time, etc., are all well characterized.

However, this is not the case in genetic logic gates, where each gate is composed of different proteins and promoters, resulting in different threshold concentration values. Furthermore, digital logic gates have the same physical quantity, i.e., voltage, as their input and output. On the contrary, genetic logic gates use different biological components including proteins, RNA, inducers, etc., to control the regulation of the corresponding output biological components. Additionally, signals in electronic circuits propagate in separate wires that do not directly interfere with each other. However, in genetic circuits, signals are molecules, drifting in the same volume of the cell, and hence easily merge with the concentration of other compounds, resulting in crosstalk with the neighboring circuit components. These facts make the timing analysis of genetic circuits very challenging.

Challenges of crosstalk have also been encountered in microelectronics; however, most of these have been solved through enhanced fabrication processes or through the development of advanced electronic design automation (EDA) tools. Similarly, advances in GDA tools may help to address these challenges, resulting in the reduction of the design complexity of genetic logic circuits.

In this chapter, a methodology is presented to perform timing and threshold value analysis of genetic logic circuit models. This methodology is implemented as a plug-in tool in DVASim. The algorithm discussed in this chapter demonstrates that it is possible to perform the timing analysis of a genetic circuit and that it can be used to

Electronic Supplementary Material The online version of this chapter (https://doi.org/10.1007/978-3-030-52355-8_4) contains supplementary material, which is available to authorized users.

achieve the desired circuit behavior. The timing analysis is performed on some of the genetic circuit models proposed in [2] and [3], and the sensitivity of circuit timings, in relation of varying different circuit parameters, is investigated. In particular, the timing sensitivity due to the *degradation rate (kd)* and the concentration of input proteins is studied.

4.1   Methodology

As mentioned in Chap. 3 that the threshold value and timing analysis can be used to verify the Boolean function of a genetic logic circuit by extracting the observed logic behavior from the simulation's results. To analyze the Boolean logic, the genetic logic circuit model can be considered as a black box. Applying all possible input combinations and observing the output can result in the combinatorial behavior of this black box. For instance, if a circuit contains two inputs, then there are four possible input combinations: 00, 01, 10, and 11.

The key challenge in determining the correct Boolean logic function from the analog simulation data is to categorize the input concentration levels into logic 0 and logic 1. As mentioned earlier, this is similar to digital electronic circuits in which a certain threshold value of input voltage differentiates logic levels 0 and 1 [1]. Digital electronic circuits are also analog in nature, but a logical abstraction has been employed to reduce the complexity of circuits. Similar abstraction has to be employed to categorize the genetic concentration levels into logic 0 and 1. To categorize these concentration levels into logic 0 and 1, the threshold value for the concentration of input proteins, which significantly affects the concentration of output protein of a genetic logic circuit, must be identified.

Different proteins in a genetic circuit may have different threshold concentration values. The approach discussed in this chapter assumes a single threshold value of the input proteins that trigger the output, instead of estimating the threshold values of each input protein separately. For instance, in the genetic AND gate (Fig. 3.5), the algorithm estimates the threshold value of LacI and TetR, which together trigger the production of GFP, rather than evaluating the separate threshold values for each of them. It may be possible that the threshold value of LacI is, for example, 13 molecules, and that of TetR is, say, 9 molecules. In this case, algorithm tells that 13 molecules is the threshold value of an entire circuit, which triggers the circuit output when the concentrations of input proteins reach this level.

Consider another example of an OR gate in which input-1 triggers the output if the molecular count is greater than 5 and input-2 triggers the output if the molecular count is greater than 10. Setting the upper input threshold to 10 would give the correct answer, i.e., the gate remains off, if the input molecular counts are (4, 7). Now, if the input molecular counts are (7, 4), then input-1 may trigger the output but it may not be considered logic 1 until the output concentration increases above 10 molecules. It is observed, through simulations that the triggered output for such scenarios is highly unstable (frequently oscillating between logic 0 and logic 1), and this region should be considered a transition region. Therefore, instead of estimating the threshold values of each input protein separately, this book focuses on the approach of estimating the global upper and lower threshold values for all inputs.

Furthermore, the approach discussed in this book considers the entire circuit as a black box and obtains the input threshold value that is required to trigger the final output. Therefore, the threshold value and the number of intermediate circuit components do not matter; the algorithm ensures that the estimated input threshold value is sufficient to trigger the intermediate circuit components all the way from input to the final output. However, the separate threshold values of intermediate circuit components can also be analyzed in DVASim (for more details, see DVASim QSG or a video demo at http://bda.compute.dtu.dk/user-manuals/d-vasim/).

4.1.1 Preliminary Analysis of Threshold Value

In order to understand the algorithm for estimating the threshold value, consider the simulation traces of the genetic AND gate generated using iBioSim [4] shown in Fig. 4.1. It shows the results from running the stochastic simulation, of the genetic AND gate for one (a) and fifty times (b) and (c). The unit of species concentration used in the circuit models of [2] is the "number of molecules." Figure 4.1a shows that both of the inputs are triggered to 10 molecules, TetR after 1000 time units and LacI after 2000 time units, and that the output is highly stochastic, which makes it difficult to determine the input threshold value. A smooth output curve is obtained by plotting the average of 50 runs, as shown in Fig. 4.1b, c.

In Fig. 4.1b, it can be observed that keeping the input concentrations to 10 molecules causes the average output concentration to stay below the level of the input concentration. Upon increasing the input concentrations further to 13 molecules, the average output concentration goes above the level of the input concentrations, as depicted in Fig. 4.1c. The same analyses can be performed with different concentration levels on different logic circuits. These analyses show a relation between the input and output proteins of a genetic circuit. On the basis of these analyses, an input–output relation of a genetic circuit can be defined in terms of its threshold value as follows:

Definition 4.1 (Threshold Value) The minimum concentration of input protein(s), which causes the average concentration of output protein to cross the concentration of input protein(s).

In the example shown in Fig. 4.1, the upper threshold value of input is 13 molecules; that is, the input concentration above 13 molecules is considered logic 1 and that below 10 molecules is considered logic 0. There is a transition region between these two levels (not shown in Fig. 4.1), where the average output concentration is not clearly distinguishable with the input concentration level. Hence, when the concentration levels of both inputs are 10 or fewer molecules i.e. logic 0; the average output concentration remains low (logic 0), else it goes high (logic 1) when the concentration of both inputs reaches 13 or more molecules (logic

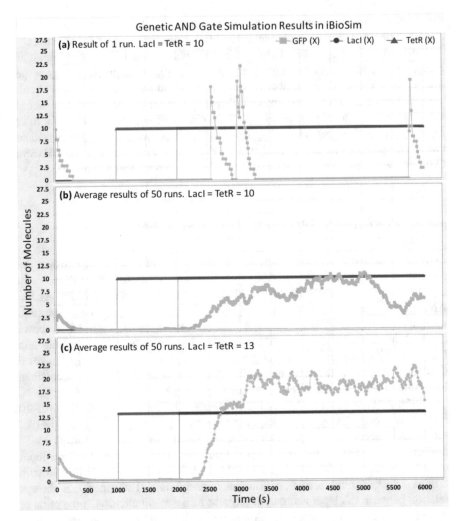

Fig. 4.1 Preliminary analysis of a threshold value for the genetic AND gate using iBioSim [4]. (**a**) Stochastic simulation results when run only once. Average results of 50 runs (**b**) showing the lower threshold value of inputs LacI = TetR = 10 and (**c**) the upper threshold value of inputs LacI = TetR = 13. Reprinted (adapted) with permission from [5]. Copyright (2017) American Chemical Society

1). This relation of input and output concentration is justified because, according to this definition, one do not need to care about how many circuit levels are cascaded between input and output. It simply identifies the input concentration required to trigger the final output. The same definition is applicable to determine the threshold values of intermediate circuit components separately.

4.1.2 Preliminary Analysis of Propagation Delay

Another important factor of a genetic circuit is its timing, i.e., how long does it take to produce the output when the input concentrations are available in a significant (threshold) amount. This is generally referred to as a *propagation delay* of a circuit in the *electronic design automation industry*.

The *propagation delay* is also an important factor to automatically obtain the correct Boolean expression from the simulation data (discussed later in Chap. 5). To understand *propagation delay*, consider Fig. 4.2 (a zoomed-in version of Fig. 4.1c), which shows that the effect of changes in the input concentration is reflected in the output concentration after a time delay of approximately 700 s. That is, the output protein takes about 700 s to cross the level of the input concentration when the inputs are triggered to their threshold value. Thus, the *propagation delay* of a genetic circuit can be defined as follows:

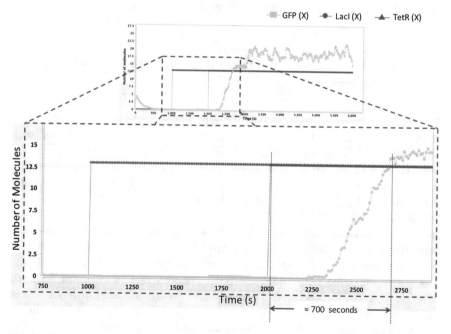

Fig. 4.2 Zoomed-in image of Fig. 4.1c indicating the preliminary propagation delay analysis using iBioSim. In this figure, the propagation delay is approximately 700 s. Reprinted (adapted) with permission from [5]. Copyright (2017) American Chemical Society

Definition 4.2 (Propagation Delay) The time from when the input concentration reaches its threshold value until the corresponding output concentration crosses the same threshold value.

Figure 4.2 shows that the output goes "high" after approximately 700 s from the time when both inputs have reached the significant concentration level (13 molecules). During these 700 s, the output remains low and hence does not produce the expected logic output. It also means that, during simulation (or even during experimentation in the laboratory), the user should not change the inputs before this time delay has elapsed.

In order to identify the threshold levels of a circuit in the laboratory, the biologist could perform this analysis by adding the input concentration periodically to see if it significantly affects the concentration of the output. To identify the input concentration, which significantly affects the output, different input combinations must be tried with different concentration levels, which is a very tedious and time consuming task to do in the laboratory. Furthermore, as mentioned above, it must be ensured that each input combination is applied after a certain time delay.

4.1.3 D-VASim Plug-in for Threshold Value and Propagation Delay Analyses

The above mentioned iterative processes can be automated by creating an algorithm to identify the threshold levels and propagation delays. We will study one such algorithm which is developed and integrated in DVASim as a plug-in tool. Since the behavior of a genetic circuit is well described by stochastic simulations, therefore the developed method is applied on the stochastic behavior of a genetic circuit obtained from the Gillespie's stochastic simulation algorithm [6, 7].

The algorithm for the threshold value and propagation delay analysis is shown as a pseudo code in Algorithm 4.1. The algorithm is initialized by some user-defined parameters as indicated in Algorithm 4.1. C_{in} specifies the value of the input protein(s) concentration, from which the tool should start its threshold analysis. The *Inc* is the value with which the input concentration is increased for each iteration, in order to observe if the resulting concentration level of input affects the concentration of the output. The C_{inE} value specifies the input concentration at which the algorithm should stop (or End) the threshold value analysis.

The algorithm also requires an initial assumption of the input–output propagation delay value, T_D. It is already mentioned earlier that the input–output propagation delay value is critical for extracting the correct logic behavior of a circuit model. Thus, it is necessary to wait until this time value has elapsed before applying the next combination of inputs. Since the time delay value is unknown for the automatic analysis, the tool begins the analysis with an assumed value and later estimates the approximate one. Assuming a higher value increases the estimation time but gives a better estimation of the threshold value. For a simulation, if every node of a genetic

Algorithm 4.1: Threshold value and propagation delay analysis. *Reprinted (adapted) with permission from [37]. Copyright (2017) American Chemical Society.*

input : C_{in}, Inc, C_{inE}, T_D, S_T, i, O_S, V_T, OC_{DUTh}, OC_{DLTh}

1 **begin**
2   **for** *all input combinations* **do**
3   **if** *(Current input concentration level (C_{inC}) == 0)* **then**
4   Determine the initial output concentration (C_{Oinit});
5   **else**
6   **while** *($C_{inC} \leq C_{inE}$)* **do**
7   **while** *(Current Time 1 (T_{C1}) $\leq T_D$)* **do**
8   *Execute simulation*
9   **if** *($C_{Os} > C_{inC}$)*** **then**
             `// ` C_{Os} ` = concentration of selected output`
                 `specie`
10             $PT = C_{inC}$ `// ` `PT = possible threshold value`
             `// Verification process`
11   **for** *(number of iterations i)* **do**
12   **while** *(Current Time 2 (T_{C2}) $\leq V_T$)* **do**
13   *Execute simulation*
14   **if** *($T_{C2} > S_T$)* **then**
15   *Trigger the input to the value of PT*
16   **end**
17   *Store the output concentration data in array*
18   **end**
19   *Take the running average of all output i arrays*
20   **end**
21   *Estimate time delay T_E and consistency OC_E*
             Terminate current while loop
22   **end**
23   **end**
24   **if** *($C_{Os} > C_{inC}$)*** **then**
25   **if** *($OC_E > OC_{DUTh}$)* **then**
26   *Consider lower threshold value = 0 if not already found Return the results and terminate all loops*
27   **else if** *($OC_E < OC_{DLTh}$)* **then**
28   *Save lower threshold level and resume analysis*
29   **else**
30   *Resume Analysis*
31   **end**
32   **end**
33        $C_{inC} = C_{inC} + Inc$
34        $T_{C1} = T_{C2} = 0$
35   **end**
36   **end**
37   **end**
    `// **Valid when ` C_{Oinit} ` is low, For high ` C_{Oinit}`, it will become`
    `// ` ($C_{Os} < C_{inC}$)
38 **end**

circuit model is not initialized to a stable value, then the output of some of the genetic circuits are initially unstable and exhibit unexpected behavior for a certain amount of time. For example, in the simulation traces (see Fig. 4.1) of genetic AND circuit (shown in Fig. 3.5), the initial values of LacI and TetR are zero, but when the simulation starts, the output, CI, of the first circuit's component (i.e. NAND gate, see Fig. 3.5) is also zero, which enables the inverter and produces GFP until the input value 0 propagate through the NAND gate.

In order to perform correct timing analysis, it is therefore required to initialize all the circuit nodes to a stable value. If the values are not initialized, it is important that the algorithm should wait for the circuit's output to become stable first. The parameter, S_T (*Settling Time*), helps the user to specify a rough value for the initial time during which the circuit's output is expected to become stable. When the algorithm performs the automatic analysis, it waits for the value defined for S_T to allow the circuit's output to become stable first and then triggers the input combinations to determine the appropriate threshold and propagation-delay values of a circuit. For small genetic circuits, containing a single gate only (for example, NOT, NAND, and NOR), and having a low degradation rate ($kd \approx 0.0015$), it is observed through simulations that these circuits usually take at least 1000 time units to become stable. This implies that, for these circuits and kd, the S_T parameter should not be less than 1000 time units. If a value less than this is chosen, then the algorithm will not be able to produce the correct estimation.

The algorithm further verifies the obtained threshold value by iterating the model for a predefined number of iterations, i. During this iterative verification process, the algorithm obtains the average propagation delay by running the model for the length of time defined by V_T for each iteration i. It also identifies the extent to which the average output for the estimated threshold value is consistent.

4.1.3.1 Illustrative Explanation of the Algorithm

In order to understand the above mentioned procedure, let us assume the parameter values shown in Table 4.1. The unit for concentration here is the "number of molecules."

Now consider the sample time scale plots shown in Fig. 4.3. To find the threshold value of the input concentration that significantly affects the output concentration, a specific input combination should be applied. This means that all the possible combinations should be checked one by one until the specific combination of inputs that triggers the output concentration is found. For logic circuits like AND, NAND, OR, NOR, and NOT, the output transition can be observed by triggering both the inputs to the same concentration level at the same time. The algorithm, therefore, triggers both the inputs combinations from 00 to 11 first, instead of following the traditional pattern of 00 $\rightarrow$ 01 $\rightarrow$ 10 $\rightarrow$ 11. Because of this, the algorithm estimates the threshold value of some circuits, for example, AND gate, relatively faster.

Table 4.1 Sample values of parameters required for threshold value and timing analyses

Parameter name	Value
C_{in}	0
Inc	2.75
C_{inE}	15
T_D	800
S_T	200
i	10
V_T	1000
OC_{DUTh}	90
OC_{DLTh}	30

Reprinted (adapted) with permission from [5]. Copyright (2017) American Chemical Society

Figure 4.3 shows how the process of automatic threshold value and timing analysis takes place by the algorithm. If more than 90% of the average output data, between instants $t2$ and $t3$, remains above the current input concentration level, that input concentration level is considered to be the upper threshold level. Similarly, if less than 30% of the average output data remains above the input level between instants $t2$ and $t3$, the input concentration level is considered to be the lower threshold level. The propagation delay is measured from the instant when the input is triggered, from its lower threshold level to its expected upper threshold level, to the instant when the average output crosses the same input level.

Figure 4.3a shows the case of input logic combination "11," i.e., when both inputs are triggered high. According to the settings shown in Table 4.1, the algorithm runs the model first by keeping the input concentration zero until the assumed time delay of 800 time units has elapsed. In order to determine if the output concentration crosses the level of input concentration as defined in Definition 4.1, or more specifically, to determine whether the output concentration goes above the input concentration level or falls below it, the initial concentration of output protein at input logic level combination "00" must be known. Therefore, during the first 800 time units (T_D), the average of the initial output concentration is obtained by keeping the concentration of both inputs zero i.e. logic 0. On the basis of this average initial output concentration, the estimation of output concentration crossing the input concentration level is performed.

Once the assumed time delay has elapsed, the input concentration level is incremented to the next level, indicated by line 33 in pseudo code Algorithm 4.1. The example case shown in Fig. 4.3a portrays the scenario of an AND gate where the initial average output of a circuit (with both input concentrations at zero) is zero. The algorithm also works for the case where the average initial output concentration is high, for instance, a NOT gate, by iteratively increasing the input concentration

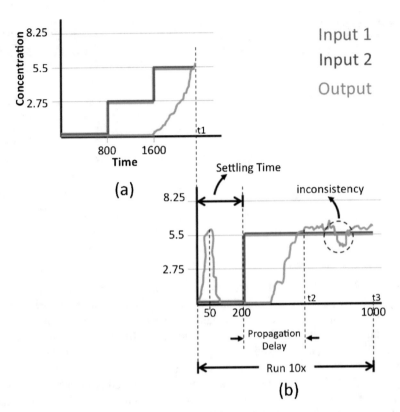

Fig. 4.3 Sample time scale plots of the genetic AND gate. (**a**) First loop to detect the threshold value. (**b**) Separate loop to verify the estimated threshold value repeatedly for predefined number of iterations, i (10 times in this case). Reprinted (adapted) with permission from [5]. Copyright (2017) American Chemical Society

and checking if the output concentration falls below the concentration level of input. It is important to note that still satisfies Definition 4.1.

Point $t1$, shown in Fig. 4.3a, implies that the algorithm halts the current loop execution when the value of the output protein crosses the input concentration level. This scenario anticipates the possible threshold value (5.5 molecules in this example) as it makes the output concentration cross the input concentration level. To verify this threshold value, the algorithm executes a separate loop to run the simulation of the circuit model for the defined number of iterations, 10 times in this example, as shown in Fig. 4.3b. This process executes in lines 11–20 in pseudo code Algorithm 4.1.

In order to measure the correct propagation delay, it is necessary to trigger the input protein, particularly from zero to the expected threshold level, only when the model's initial output is settled. As mentioned before, the outputs of some circuits are unexpectedly high which gradually settles down to zero. This scenario is depicted in Fig. 4.3b. Therefore, the initial concentrations of input proteins must not be triggered to their expected threshold level until the output becomes stable. As mentioned above, the parameter Settling Time, S_T, allows a user to provide a period of time by which the initial output is expected to become stable. This is the time at (or after) which the algorithm triggers the inputs, to their expected threshold level, to determine the time it takes to trigger the output concentration. If a low value is assumed for S_T, the algorithm may produce incorrect propagation delay. For example, in Fig. 4.3b, if a value 50 would be chosen as a settling time, the inputs would be triggered at 50 time units. At this instant, when the inputs are triggered to their expected threshold level, the concentration of output is already above the threshold level and thus the algorithm would end up estimating the propagation delay value to zero.

 Depending on the complexity of a circuit and the degradation rate (kd), the value of S_T should be chosen carefully, otherwise the algorithm will not produce the correct results.

The simulation output data from all 10 iterations are averaged to obtain the average estimated propagation delay and the inconsistency present in the output plot for the estimated threshold values. The inconsistency, illustrated in Fig. 4.3b, is calculated by determining the size of the average output data, which is less than the input concentration level immediately after the output crosses the input level for the first time, i.e., the inconsistency is estimated between points $t2$ and $t3$ as shown in Fig. 4.3b. In other words, for examining the upper threshold level, the idea is to determine how consistently the average output data remains above the input concentration level between points $t2$ and $t3$. The algorithm accepts the estimated threshold value based on the user-defined parameter, *% acceptance of consistency*, shown as OC_{DUTh} (for upper threshold level) and OC_{DLTh} (for lower threshold level) in pseudo code Algorithm 4.1. The results are accepted if the estimated consistency is greater (for upper threshold) and less (for lower threshold) than the user-defined values, OC_{DUTh} and OC_{DLTh}, respectively. This is shown in the lines 25–30 in pseudo code Algorithm 4.1. The results are otherwise discarded and the algorithm resumes the analysis from point $t1$, shown in Fig. 4.3a. The percentage output consistency is calculated according to Eq. 4.1.

$$\% \ output \ consistency = \frac{O_{t2-t3} - D}{O_{t2-t3}} \tag{4.1}$$

where:

O_{t2-t3} = Size of the average output data between instants $t2$ and $t3$ (Fig. 4.3b).
D = Deviation, which defines the number of times the output data is found to be deviated (greater or less) from the expected threshold value.

The quantity D in Eq. 4.1 is considered to be different in two different cases, i.e., when the initial input concentration is found low, then D in Eq. 4.1 indicates the number of times the output data is found "less" than the threshold value, as in the case shown in Fig. 4.3. Else, if the initial input concentration is high, then D signifies the number of times the output data is found "greater" than the threshold value. For the sample parameters (Table 4.1) used for the sample plots shown in Fig. 4.3, the algorithm estimates the input concentration as the upper threshold level if the output consistency is 90% or above. Likewise, the input level is assessed as the lower threshold level if the estimated output consistency is less than 30%.

4.2   Experimentation by Simulation

Now we can use the developed algorithm for threshold and timing analysis of different genetic circuit models. In this chapter, we will discuss the results of ten genetic circuit models shown in Fig. 4.4—nine are obtained from [2] and one from [3]. The genetic implementation and the description of these circuits can be found in [2] and [3], respectively. These circuits are considered fairly complex in the context of genetic circuits, because each gate is composed of several genetic components. Their kinetic interactions are described by a number of mathematical equations in the SBML model. The SBML models of these genetic logic circuits are run on DVASim and their threshold value and propagation delay analyses are performed.

In microelectronic devices, the behavior of a circuit depends on many different parameters. For example, in MOS transistors, the drain current depends on the width and length of gate, oxide capacitance, gate-to-source voltage, etc. [8]. Similarly, the behavior of a genetic circuit also depends on different parameters, including degradation rate, forward repression binding rate, forward activation binding rate, etc. These parameters of a genetic circuit model are described in the SBML file. We will discuss the effects of varying the degradation rate (kd) on the propagation delay and the threshold value of a circuit.

The degradation rate is the rate at which a chemical compound (e.g. a protein) is decomposed into intermediate products, i.e., a produced protein will be effective only for a certain period of time determined by the degradation rate. A zero degradation rate means that the protein does not degrade and hence will be effective forever. This is an often-used assumption, which clearly is not realistic, which is why understanding the impact of the degradation rate on the timing analysis is an important investigation.

Since majority of the genetic circuit models we are studying are presented in [2] and the value of degradation rate used in these models is 0.0075, therefore the effect of varying degradation rate, around the default value, on the timings of genetic circuits is presented here. Five different values of degradation rate kd (0.0015, $0.0055, 0.0095, 0.0135, 0.0215$) are chosen to study its impact on the behavior of a genetic circuit. It has been experimentally observed that the variation in degradation rate (kd) greatly effects the settling time, S_T, of an output. Hence, for each circuit, different parameter values (shown in Table 4.1) are chosen, except the number of iterations, i, and *% acceptance of consistency* for upper and lower threshold values

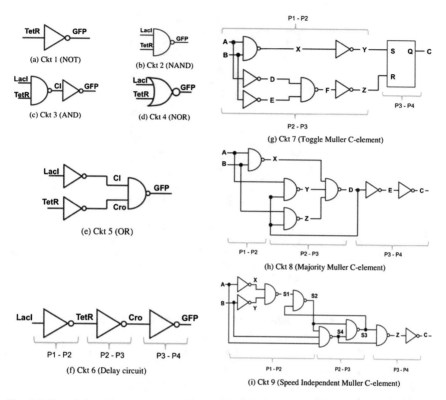

Fig. 4.4 Experimental genetic circuits, obtained from [2], for timing analysis. (**a**–**e**) Schematics of standard digital logic gates, whereas the gates (**c**) and (**e**) are implemented with NOT and NAND logic. More complex circuits, (**f**–**i**), are further categorized into three intermediate levels P1–P2, P2–P3, and P3–P4. The timing analyses are performed on these three levels separately, which are mentioned in Table 4.2. The SR latch shown in Ckt 7 is asynchronous and do not require a clock input. Reprinted (adapted) with permission from [5]. Copyright (2017) American Chemical Society

(OC_{DUTh} and OC_{DLTh}), which were set to 5, 70, and 30, respectively, for all circuits. The value of $OC_{DUTh} = 70\%$ is purposely chosen in order to demonstrate that it is affected by threshold values.

Figure 4.5 shows how DVASim reports the outcomes of a threshold value and propagation delay analysis once the algorithm finishes execution. This figure shows the threshold value and timing analysis results obtained for Ckt 8 (Fig. 4.4h) when degradation rate (*kd*) was set to *0.0135*. It indicates that the estimated upper and lower threshold values are 6.5 and 3.25 molecules with 98.4 and 25.5% consistency, respectively. It also calculates the approximate input–output propagation delay value to 620 time units with a standard deviation of ±190.08, in this case based on five iterations. When the results are obtained, the user may interact with the model during run-time, apply all the possible input combinations in a significant amount, and match the propagation delay with the one estimated by DVASim.

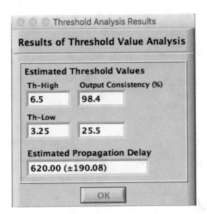

Fig. 4.5 Results of threshold value and propagation delay analysis of Ckt 8 generated by D-VASim for $kd = 0.0135$. The estimated upper and lower threshold values are 6.5 and 3.25 molecules with 98.4 and 25.5% consistency, respectively. The approximate input–output propagation delay value is 620 time units with ±190.08 standard deviation. Reprinted (adapted) with permission from [5]. Copyright (2017) American Chemical Society

Figure 4.6 shows the simulation traces of the same circuit (Ckt 8) with $kd = 0.0135$. To keep this chapter concise, only the screen shots of the analysis results and graphical simulation of Ckt 8 (for $kd = 0.0135$) are included. In Figure 4.6, the circuit's inputs are A and B; and the output is C. It can be observed that the initial concentration of output protein, C (shown as green plots), is high above the threshold value and it takes approximately 400 time units to settle down. Furthermore, when the input concentrations are triggered to their lower threshold level, i.e., 3.25 molecules, the output concentration remains zero. When the input concentration levels are triggered sharply to their estimated threshold value i.e. 6.5 molecules, the output of a circuit triggers high approximately after 800 time units.

 DVASim allows a user to analyze analog as well as digital waveforms of genetic circuit models. See DVASim QSG for more information.

4.2.1   Effects of Varying kd on the Threshold Values and Propagation Delays

Figure 4.7 shows the graphical plots of timing analysis of all ten circuits. The values of propagation delays are plotted along the y axis on the left-hand side. The threshold values and percentage of output consistency for each value of kd is plotted along the y axis on the right-hand side. The x axis contains the degradation rate values. The general impression of these experiments is that the propagation delay of genetic circuit decreases with the increase in degradation rate (kd). This is because when

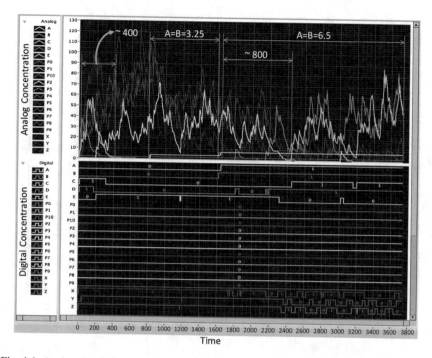

Fig. 4.6 Analog simulation traces of Ckt 8 with its corresponding digital waveforms for $kd = 0.0135$. Reprinted (adapted) with permission from [5]. Copyright (2017) American Chemical Society

the degradation rate is high, the protein degrades faster and thus contributes in the reduction of propagation delay. However, the propagation delay does not seem to have an inverse relation with the degradation rate. The propagation delay for all circuits dropped considerably with the first decrement of 40×10^{-4} in kd; and then it decreases slowly for the next higher values of kd.

The standard deviation in propagation delays, calculated for the specified number of iterations (i.e. five in these experiments), is also included for each circuit in the plots shown in Fig. 4.7. It can be noticed that the propagation delay of a circuit is more variable for low degradation rates. The variation in the propagation delay decreases with the increase in degradation rate; however, a high degradation rate makes the cascaded circuit's output unstable. This is because the genetic components decay quickly when the degradation rate is high, thus causing the circuit's logic to switch faster even when a small input concentration is applied. This also reduces the transition region between the upper and lower threshold levels, as shown in the data of Ckt 8 in Fig. 4.7. However, it can also be noticed for Ckt 8 that a transition region is small for $kd = 0.0135$ (i.e. 3.25) as compared to $kd = 0.0215$ (i.e. 6.5). This is because, at $kd = 0.0215$, Ckt 8 becomes unstable and produces glitches of high output even when the input concentration levels were kept to zero. This is the reason why the lower threshold value of Ckt 8 at $kd = 0.0215$

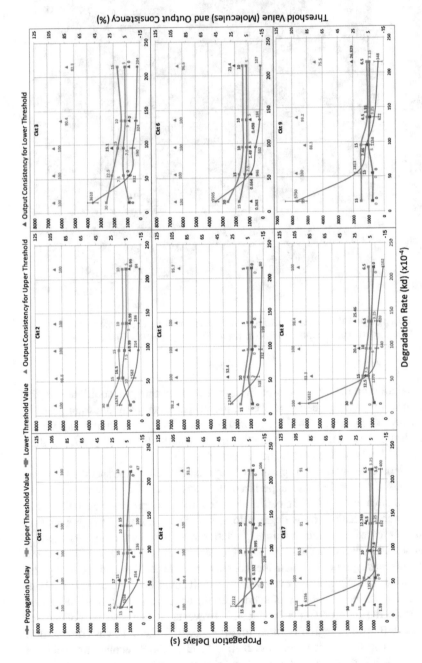

Fig. 4.7 Effects of varying degradation rate (kd) on the propagation delay of genetic logic circuits. Reprinted (adapted) with permission from [5]. Copyright (2017) American Chemical Society

is estimated to be zero. It has been observed that the lower threshold value of all circuits approaches zero with the increase in *kd*.

4.2.2 Effects of Varying Threshold Values on the Propagation Delays

Beside these analyses, some other interesting facts of varying the upper threshold value on the propagation delay of a circuit can be observed. It can be noticed that the smaller concentrations of input protein have a weak impact on the output protein, which is analogous to the behavior of microelectronic devices. For instance, in MOS transistors, weak applied V_{GS} (gate-to-source voltage) results in the weak drain current, I_D [8]. This effect can be observed in the graphical plot of Ckt 4 in Fig. 4.7. In this plot, for $kd = 0.0215$, the threshold value of a circuit is reduced to 5 molecules as compared to its previous data point, which is 10 molecules at $kd = 0.0135$. Due to the increment in *kd*, the propagation delay at this point is supposed to decrease if the input threshold value remains the same. However, it slightly increases because the input threshold is reduced to 5 molecules. This effect has been observed on other circuits as well during the run-time simulation. This inverse relation between propagation delay and threshold value holds true to a certain extent, and then further increment or decrement in the applied input concentration does not affect the propagation delay.

4.2.3 Effects of Varying Threshold Values on the % Output Consistency at High kd

For higher values of kd, the output consistency of the upper threshold level is increased by reducing the threshold value. This is shown in the plots of Ckt 9 in Fig. 4.7. The output consistency of the upper threshold level, at $kd = 0.0135$, was reduced to 49% (not shown in Fig. 4.7) when the threshold value was set to 30 molecules. Then the output consistency of Ckt 9 was analyzed, at $kd = 0.0215$, by keeping the threshold value to the same level, i.e., 30 molecules (not shown in Fig. 4.7), and noticed that the output consistency was decreased to 2%. Then the threshold value was decreased to 6.5 molecules and the output consistency was increased to 75.5%, as shown in Fig. 4.7.

4.2.4   Other Parameters Effecting the Threshold Values

The values of upper and lower threshold levels also depend on the parameter, *Inc* (see Table 4.1), which specifies the input concentration to be added to the previous input concentration level during each iteration, *i*. For example, in the case of Ckt 2 and Ckt3, the value of *Inc* was set to 30 at $kd = 0.0015$. The algorithm thus triggers the input concentration from 0 to 30 directly during the analysis. Because of this, the average output was found to be 100% consistent for upper threshold level, which results in estimations of the upper and lower threshold levels of 30 and 0 molecules, respectively. If a lower value of *Inc* would be chosen, the results would be different but more precise.

4.2.5   Intermediate Propagation Delays

The intermediate delays of larger genetic circuit models are also affected by the variations in *kd*. The intermediate delays can be analyzed by splitting the circuits into three points of measurements, as shown in Fig. 4.4. The propagation delays for each of these points are mentioned in Table 4.2. The propagation delay, indicated by a point of measurement *P1–P4* in Table 4.2, is the entire circuit propagation delay. The reader should not confuse these estimations with those depicted in Fig. 4.7. The results mentioned in Fig. 4.7 are estimated by DVASim using the proposed algorithm; and the results mentioned in Table 4.2 are those that are obtained by a user through a run-time interactive stochastic simulation.

   The argument that a circuit's output becomes unstable for larger values of *kd* can also be supported by observing the intermediate delays of Ckt 6 for $kd = 0.0215$ in Table 4.2. As shown in Fig. 4.4, Ckt 6 is composed of three inverters connected back-to-back in series. When input protein LacI is triggered to its threshold value, it suppresses the production of TetR. When the concentration of TetR drops below its threshold level, it produces Cro, which in turn suppresses the production of output protein, GFP. However, the intermediate propagation delays of Ckt 6 for $kd = 0.0215$ show that when the input protein, LacI, is triggered to the estimated threshold value, the overall output of a circuit, GFP, is produced in 56.5 time units. However, one of the intermediate outputs, Cro, is produced in a significant amount after ≈ 70 time units, which is greater than the propagation delay of the entire circuit. This invalidates the desired circuit's behavior and makes the output unstable, which indicates that the circuit does not behave as designed.

Table 4.2 Intermediate propagation delays of genetic logic circuits

kd ($\times 10^{-4}$)	Points of measurement	Propagation delays (s)			
		Ckt 6	Ckt 7	Ckt 8	Ckt 9
15	P1–P2	1763.12	3243	1863	4379
	P2–P3	1067.91	4764	1237	3854
	P3–P4	554.53	1140	2500	1292
	P1–P4	3350	5904	5500	6463
55	P1–P2	382	603	394	936
	P2–P3	193.7	973	425	395
	P3–P4	246.4	410	405	171
	P1–P4	805	1400	1224	1646
95	P1–P2	80	340	122	441
	P2–P3	100	380	235	53.045
	P3–P4	180	280	235	64
	P1–P4	360	664	631	1003
135	P1–P2	83	230	39	311
	P2–P3	81	265	253	.112
	P3–P4	54	240	279	330
	P1–P4	215	506	573	878
215	P1–P2	43	229	70	9
	P2–P3	70	173	103	120
	P3–P4	58	47	37	81.63
	P1–P4	56.5	267.75	208	400

Reprinted (adapted) with permission from [5]. Copyright (2017) American Chemical Society

4.2.6 Experimentation on the SBML Model of Real Genetic Circuit

As mentioned earlier, we have included the SBML model of one of the genetic circuit which has actually been tested in wet lab. The genetic AND gate circuit (composed of inverters and NOR gates) from [3] is chosen for the experimentation in DVASim. The genetic circuits presented in [3] were first developed on a tool named *Cello*, which generates the SBOL file. Unlike SBML, the SBOL representation does not describe the behavior of a biological model. Therefore, the SBOL-SBML converter [9] is used to generate the behavioral model of the above mentioned real genetic AND circuit. This SBOL-SBML converter is available as a plug-in tool in iBioSim [4], which uses the default parameters while defining the reaction kinetics during the conversion process.

Since, the actual parameters, like degradation rate, forward repression binding rate, etc., are not disclosed in [3], therefore the default iBioSim parameters are used to perform the timing analysis of real genetic AND gate circuit. However, the parameter values can always be changed, and new parameters can also be added to observe more realistic results.

The SBOL file generated by Cello does not include the input sensor block of a
circuit (which includes the input inducers); thus these inducers are also not included
during the SBOL-SBML conversion process.

The input inducers are added manually in the SBML model using iBioSim,
as shown in Fig. 4.8. The components inside the yellow-dashed box are manually
added, and rest of the model is a result of SBOL-SBML conversion processs-SBML
conversion. In this figure, it is shown that when both of the input inducers, aTc
and IPTG, are present, they form a complex with their corresponding regulators,
TetR and *LacI*, respectively. These regulators then gradually stop inhibiting their
respective promoters, which eventually leads to the production of the output, yellow
fluorescent protein (*YFP*).

Figure 4.9 shows the timing analysis results of the SBML model of the genetic
AND gate circuit [3]. All these analyses for different degradation rates were
obtained within 30 min, and the simulations with all possible input combinations
were performed within 10 min. This is obviously faster compared to testing the
model in a lab, where the models were first placed in the logic-0 state for 3 h and
then switched to other possible states, one by one, each for another 5 h [3].

Figure 4.9 indicates that the results of the genetic AND gate [3] are similar
to those obtained for the other 9 genetic circuit models [2]. In general, it is
observed that the propagation delay, threshold value, and the degradation rate are
all interlinked. The output of a circuit is stable for small values of kd but it increases
the propagation delay. The variation in the propagation delay is also greater for small
values of kd. On the other hand, the output of a circuit becomes unstable for large
values of kd but decreases the propagation delay. Large values of kd also contribute
to a reduction of threshold value to a certain point. This is because the circuit
becomes faster for large kd; therefore, a small input concentration is sufficient to
trigger the output protein. The degradation rate cannot be increased beyond a certain
point, because it makes the output highly oscillating. This implies that the threshold

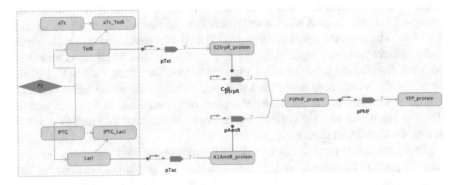

Fig. 4.8 SBML design of the genetic AND gate circuit obtained from [3]. Reprinted (adapted)
with permission from [5]. Copyright (2017) American Chemical Society

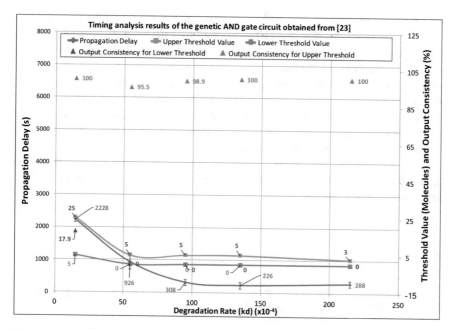

Fig. 4.9 The effects of varying degradation rate (*kd*) on the propagation delay and threshold value of the genetic AND gate circuit [3]. The propagation delay of a circuit decreases with the increase in *kd*. Also, low input concentrations are required to trigger the output of a circuit at higher values of *kd*. Note: These results may differ when the actual parameter values are used

value of a circuit cannot be decreased beyond a certain point. This corresponds to the scaling trends for the MOSFET device, where the gate width cannot be reduced beyond a certain number of nanometers.

4.3 Summary

In this chapter, a methodology to perform the timing analysis of genetic logic circuits is studied, which is implemented and tested in DVASim. The threshold value and timing analysis are performed primarily on entire circuits instead of on each individual circuit component. However, DVASim is also able to analyze the threshold value and timing analysis of individual circuit components. In this chapter, DVASim is shown to estimate the overall threshold value of an entire circuit, which gives user a minimum value of input species required to trigger the output of a genetic circuit.

The effects of circuits' timings upon varying certain parameters are also explored. This may assist genetic circuit designers in finding an appropriate set of parameters to achieve the desired timings of a genetic circuit. D-VASim could actually help reduce the time consuming in-vitro experiments (laboratory experiments) to analyze and design genetic circuits with desired behavior and timings. We anticipate that the ability of analyzing the timings of genetic circuit may open up a new research area, which may help biologists and scientists to design and characterize the timing properties of genetic circuits. Depending on the complexity of a genetic circuit and the user-defined settings for these analyses, D-VASim may take up to an hour to estimate the threshold value and propagation delays. This estimation time is still reasonable as compared to the number of days of laboratory experimentation, which are required only for a single combination of inputs with a specific set of parameters.

This methodology of threshold value and propagation delay analyses is used in the next chapter to perform the experimentation and logic analysis on some of the real genetic circuit models [3].

Problems

The following problems are based on the SBML models of genetic logic circuits (shown in Fig. 4.4) which are enclosed in ~/Resources/Problems/Chapter4/ Models/ directory.

 You would need to load a circuit model in main DVASim first, then "**Generate SSA VI**" to trigger the virtual simulation environment for stochastic simulations. See DVASim QSG for more details.

 Make sure to turn the **Threshold Value Analysis** switch to **Auto** mode, before pressing the **Run** button.

4.1 For the following circuit models, execute the automated threshold value and timing analysis for the specified value of *Kd* and verify the results using the interactive stochastic simulation.

(a) *Ckt1_not_gate*; $Kd = 0.0015$
(b) *Ckt2_nand_gate*; $Kd = 0.0055$
(c) *Ckt3_and_gate*; $Kd = 0.0095$
(d) *Ckt4_nor_gate*; $Kd = 0.00135$
(e) *Ckt5_or_gate*; $Kd = 0.00215$

4.2 For all the models of *Ckt6 through Ckt9*, using interactive stochastic simulation, perform the propagation delay analysis of intermediate points as shown in Fig. 4.4 and fill up the following table. Compare your results with the ones shown in Table 4.2.

Note Because of the stochasticity, your results might slightly differ with those presented in Table 4.2.

kd ($\times 10^{-4}$)	Points of measurement	Propagation delays			
		Ckt 6	Ckt 7	Ckt 8	Ckt 9
15	P1–P2				
	P2–P3				
	P3–P4				
	P1–P4				
55	P1–P2				
	P2–P3				
	P3–P4				
	P1–P4				
95	P1–P2				
	P2–P3				
	P3–P4				
	P1–P4				

4.3 Verify that the % output consistency of a circuit is increased by reducing the threshold value at higher values of Kd. Use the model of *Ckt 9* as suggested in the Sect. 4.2.3. Generate the results similar to the one shown in Fig. 4.5

4.4 Using Ckt 7, verify that increasing the input concentration may also result in the reduction of propagation delay. Use the value of $kd = 0.0015$. Also, in the threshold value analysis settings, first use the minimum input transition value (*Inc*, as shown in Table 4.1) to 15 (i.e. *Increment of = 15*, see DVASim QSG). To analyze for higher concentration values, use 60 for the same parameter.

Since the value of *Kd* is low, the analysis may take longer. Also, the value of Settling Time (S_T) should be chosen as high as possible to make sure the initially unstable output settles down properly.

Hint: Fig. 4.7 can be used to seek help in setting up the appropriate parameters for threshold/timing analysis.

4.5 For Ckt 10, perform the automated threshold value analyses for all the five values of *Kd* used in this chapter. Also, perform the interactive stochastic simulations to verify the results obtained by an automated procedure.

References

1. B. Mark, in *Complete Digital Design: A comprehensive guide to digital electronic and computer system architecture* (McGraw-Hill, New York, 2003)
2. C.J. Myers, *Engineering Genetic Circuits* (Chapman and Hall/CRC, New York, 2009)

3. A.A.K. Nielsen, B.S. Der, J. Shin, P. Vaidyanathan, V. Paralanov, E.A. Strychalski, D. Ross, D. Densmore, C.A. Voigt, Genetic circuit design automation. Science **352**(6281), aac7341–aac7341 (2016)
4. C.J. Myers, N. Barker, K. Jones, H. Kuwahara, C. Madsen, N.P.D. Nguyen, iBioSim: a tool for the analysis and design of genetic circuits. Bioinformatics **25**(21), 2848–2849 (2009)
5. H. Baig, J. Madsen, Simulation approach for timing analysis of genetic logic circuits. ACS Synth. Biol. **2**, acssynbio.6b00296 (2017)
6. D.T. Gillespie, A general method for numerically simulating the stochastic time evolution of coupled chemical reactions. J. Comput. Phys. **22**(4), 403–434 (1976)
7. D.T. Gillespie, Exact stochastic simulation of coupled chemical reactions. J. Phys. Chem. **81**(25), 2340–2361 (1977)
8. R.R.K.C.R.P. P. Visveswara, R.B. Bhaskara, M.K. Rama, Electronic devices and circuits, in *Pearson Education*, 2nd edn. (2007)
9. N. Roehner, Z. Zhang, T. Nguyen, C.J. Myers, Generating systems biology markup language models from the synthetic biology open language. ACS Synth. Biol. **4**(8), 873–879 (2014)

Chapter 5
Genetic Circuits Logic Analysis

After obtaining the *threshold value* and *propagation delay* of a genetic circuit through the procedure defined in Chap. 4, a user can begin experimenting on the genetic circuit model and apply all the possible input combinations. Once all the possible input combinations of a genetic circuit are applied, the experimental data (stochastic simulation data) can then be used to obtain (or validate) the logical behavior of a genetic circuit model.

In this chapter, the logic analysis and validation algorithm is presented, which extracts the logic behavior from the simulations and provides a fitness value that can be used to infer how likely it is that the circuit will actually work after implementation in the laboratory. The algorithm discussed in this chapter is scalable and able to analyze n-input genetic logic circuits. The logic analysis of genetic circuits is useful in two ways—first, it allows the user to verify more complex genetic logic circuits, build by cascading several genetic logic gates; secondly, it helps the user to extract the Boolean logic of a circuit even when the user does not have any prior knowledge about its expected behavior. Similar to *timing analyzer* plug-in, the methodology for logic analysis is also available as a plug-in tool in D-VASim. This approach has been tested on 15 different genetic circuit models, with different level of complexity, obtained from [3] and [1].

5.1 Methodology

As discussed in Chap. 4, *threshold value* and *propagation delay* of I/O species are two important parameters required to obtain a correct Boolean expression of a genetic circuit. Definition 4.1 states that the *threshold value* defines a significant

Electronic Supplementary Material The online version of this chapter (https://doi.org/10.1007/978-3-030-52355-8_5) contains supplementary material, which is available to authorized users.

amount of concentration, which categorizes the analog concentrations into digital logics 0 and 1. Also, Definition 4.2 specifies that the *propagation delay* is the time required to reflect the changes in input species concentrations on the concentration of output species.

During the experimentation, if the input species concentrations are applied below their threshold levels and each of the input combination is changed before the propagation delay has elapsed, then the circuit never produces a correct output for some of the input combinations. We first use the timing analysis plug-in of D-VASim [4], discussed in Chap. 4, to obtain the threshold value and the propagation delay of a circuit. These results are then used to perform experiments on genetic circuit models and log all experimental simulation data. The simulation data is then given to the proposed algorithm to extract the logical behavior of a circuit.

5.1.1 Overview

Algorithm 5.1 shows the pseudo code of the main procedure of the logic analysis and verification algorithm, which contains three sub-procedures: **CaseAnalyzer, VariationAnalyzer**, and **ConsBoolExpr**, discussed separately in the Algorithms 5.2, 5.3, and 5.4, respectively. Some initial parameters (N, SD_{An}, Th_{VAL}, FOV_{UD}, I_S, and O_S) are required to execute the algorithm, where N corresponds to the total

---

Algorithm 5.1: The main procedure of logic analysis and verification algorithm. © [2017] IEEE. Reprinted, with permission, from [7]

---
1 **begin**
2   **INITIALIZE** (N, I_S, O_S, SD_{An}, Th_{VAL}, FOV_{UD});
3   SD_{size} = Calculate the size of analog simulation data, SD_{An}
4   SD_{Dig} = **ADC** (N, SD_{An}, SD_{size}, Th_{VAL})
      `// SD_Dig = digital simulation data`
5   (*nc, Case_O, Case_I*) = **CaseAnalyzer** (N, SD_{size}, SD_{Dig})
      `// Case_O = Array holds the output values for each input`
      `   combination`
      `// Case_I = Array holds the number of occurrences of each`
      `   input combination`
      `// nc = total number of possible input cases (or`
      `   combinations): 2^N`
6   (*O_Var, HIGH_O*) = **VariationAnalyzer** (*nc*, SD_{Dig}, *Case_O*)
      `// O_Var = Array to monitor variations in the output for`
      `   each case, nc`
      `// High_O = Array to hold the number of times the output`
      `   is high for each case, nc`
7   (*BoolExpres, PFoBE*) = **ConsBoolExpr** (*O_Var, Case_I, HIGH_O, nc, N, FOV_UD*)
      `// BoolExpres = Contains the estimated Boolean expression`
      `// PFoBE = Specifies the percentage fitness of estimated`
      `   Boolean expression in the simulation data`
8 **end**

---

number of input species, SD_{An} refers to the simulation data of all I/O species, Th_{VAL} denotes the upper threshold value of I/O species, FOV_{UD} is the user-defined percentage of acceptable variation in the output data (described later), and I_S and O_S specify the names of input and output species, respectively. By giving users an ability to select the input and output species, they can perform Boolean logic analysis on the entire circuit as well as on the intermediate circuit components.

In the simulation of electronic circuits, a logical abstraction is typically applied in which it is only considered if the wire is in high or low state, instead of tracking the exact voltage value. In order to utilize a similar abstraction level here, the algorithm first converts the analog simulation data into digital data with the help of upper threshold values extracted from the *timing analyzer* plug-in of D-VASim (see Chap. 4). This step is shown as the sub-procedure **ADC** at line 4 in Algorithm 5.1. The algorithm scans the chosen N input and an output species and converts their analog values into digital values, based on the upper threshold value provided. Once the analog data is converted to logic high and low, the exact concentration of proteins is no longer needed in order to obtain the Boolean logic of a genetic circuit.

Analog simulation data of species is first converted to Binary data based on the concentration level greater or lesser than the threshold value.

5.1.2 Input Combinations Analysis

The response time of a genetic circuit is important in order to obtain the correct behavior. Therefore, each input combination has to be applied for sufficient amount of time to observe its correct response on the output species. In electronic circuits, the signals propagate in separate wires and applied voltage remains constant. However, the signals in genetic circuits are molecules drifting in the same volume of a cell and easily merge with the concentrations of other compounds. Due to this, the concentrations of a species in a genetic circuit always vary, and may go up or down below the threshold level at any instant of time. Because of this unstable behavior, for each input combination, it is required to obtain continuous binary streams of output species to extract the correct behavior of a genetic circuit.

The next sub-procedure, **CaseAnalyzer**, shown in Algorithm 5.2 analyzes the number of times each input combination occurs (line 8) and logs their corresponding output binary data streams (line 9–12). In order to understand this procedure, consider the sample simulation plots in Fig. 5.1a, which are produced from the 2-input genetic AND gate of Fig. 3.5. **CaseAnalyzer** processes the data and generates output as depicted in the first three columns in Fig. 5.1b. This data express, for each input combination, the number of simulated data points as well as the output digital data stream of logic-0 and 1 converted according to the upper threshold levels. In this example, the case of input combination 00 appears about 1850 times in total.

Algorithm 5.2: Pseudo code of the procedure **CaseAnalyzer**

---

    input : N, SD_{size}, SD_{Dig}
    output: nc, Case_O, Case_I
1 **begin**
2       $nc = 2^N$;
3   *Set Array* Case_I[nc] = 0;
4   *Set Array* Case_O[nc][SD_{size}] = 0;
5   *Set icv*; // to read Input Case Values
6   **for** *all* $j \in SD_{Dig}$ **do**
7            $icv = At$ j^{th} *value, read the value of corresponding inputs' combination*;
8   Case_I [icv] = Case_I [icv] + 1;
9   **if** (j^{th} *value of* SD_{Dig} *output-specie for input case icv == 1*) **then**
10   *Set* Case_O [icv][j]) = 1;
                // Note that the Output is High for specific case
                   icv at simulation instant j
11   **else**
12   *Set* Case_O [icv][j]) = 0;
                // Note that the Output is Low for specific case *icv*
                   at simulation instant j
13   **end**
14   **end**
15 **end**

The small glitch between 4650 and 6350 time units indicates the stochastic nature of the model.

It shows that the logic-0 of GFP may refer to a concentration which is less than its threshold value but may not be sharply zero. Also, the output of some genetic circuit models is initially high which gradually reduces down to zero, as shown in Fig. 5.1a. These unwanted high peaks should be filtered out to obtain the correct Boolean expression.

For each input combination, the corresponding data stream of the output species is also extracted, as shown in the third column of the table shown in Fig. 5.1b. In this example, the output data stream contains binary 1's for two input combinations—00 and 11. In this case, it is already known that the output of a circuit is initially high, when both of the inputs are low, and settle down to zero gradually. Furthermore, Fig. 5.1a depicts a short period of time in which the output oscillates around the upper threshold value (between 6350 and 9400 time units), before entering into a stable logic-1 state. This happens when both inputs are triggered high (i.e. 11). In order to examine such scenarios, the digital output data streams, corresponding to each input combination, are analyzed for stability through the sub-procedure, **VariationAnalyzer**, (line 6 in Algorithm 5.1).

5.1.3 Variation Analysis and Boolean Expression Construction

The pseudo code of the sub-procedure **VariationAnalzyer** is shown in Algorithm 5.3. **VariationAnalzyer** examines the output data stream (lines 8–27) and counts how many times the output oscillates (or varies) between logic-1 and 0. It first calculates the number of times a logic-1 appears for a specific input combination (line 19). In the example shown in Fig. 5.1b, the logic-1 appears for 3 and 1875 times for the input combinations 00 and 11, respectively. It then analyses for each of these input combinations changing 0-to-1 and 1-to-0 (i.e. how many times the output varies). In Fig. 5.1b this happens twice for input combination 00 and 7 times for 11. Since the output is high when both the inputs are the same, one may end up estimating the logical behavior of this circuit to be an XNOR gate (=> output is High when inputs are same) if the simulation data is not filtered out correctly.

To obtain the correct Boolean expression, two filtrations of the data are performed by the sub-procedure, **ConstBoolExpr** (line 7 in Algorithm 5.1). The first one is the estimation of fraction of variation (FOV_{EST}) according to Eq. 5.1:

$$FOV_{EST_i} = \frac{O\_Var[i]}{Case\_I[i]} \qquad (5.1)$$

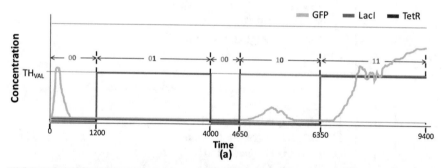

Inputs (i)	Input occurrences (Case_I[i])	Output stream for each input combination (Case_O[i])	Output occurrences (HIGH_O[i])	Unstability (O_Var[i])
00	1850	00011100...................00000	3	2
01	2800	000.............................00000	0	0
10	1800	000.............................00000	0	0
11	3050	00..011...10011010001...11	1875	7

(b)

Fig. 5.1 Logic analysis and verification process. (**a**) Sample plots of 2-input genetic AND gate. (**b**) Sample data for illustrating the input case and variation analysis. © [2017] IEEE. Reprinted, with permission, from [7]

Algorithm 5.3: Pseudo code of the procedure **VariationAnalyzer**

---

input : *nc, Case_O*
output: *O_Var, HIGH_O*
1 **begin**
2 | *Set* Prev_OP_State = 0; // Previous digital state of output
   | specie
3 | *Set* Curr_OP_State = 0; // Current digital state of output
   | specie
4 | *Set* O_Var [*nc*] = 0;
5 | *Set* HIGH_O [*nc*] = 0;
   | // loop through all input cases nc
6 | **for** *all i ∈ nc* **do**
   | // loop through digital data of output species
7 | **for** *all j ∈ Case_O* **do**
8 | **if** *(j^{th} value of Case_O for input case i == 0)* **then**
9 | **if** *(Prev_OP_State == 1)* **then**
10 | *Increase* O_Var [*i*] (i.e. for i^{th} input case) *by* 1;
11 | *Set* Curr_OP_State = 0;
12 | *Set* Prev_OP_State = Curr_OP_State;
13 | **else**
14 | *Keep* O_Var [*i*] *to its previous value*;
15 | *Set* Curr_OP_State = 0;
16 | *Set* Prev_OP_State = Curr_OP_State;
17 | **end**
18 | **else**
   | // Count number of times the output is high for
   | i^{th} input case
19 | HIGH_O [*i*] = HIGH_O [*i*] + 1;
20 | **if** *(Prev_OP_State == 0)* **then**
21 | *Increase* O_Var *by* 1;
22 | *Set* Curr_OP_State = 1;
23 | *Set* Prev_OP_State = Curr_OP_State;
24 | **else**
25 | *Keep* O_Var [*i*] *to its previous value*;
26 | *Set* Curr_OP_State = 1;
27 | *Set* Prev_OP_State = Curr_OP_State;
28 | **end**
29 | **end**
30 | **end**
31 | **end**
32 **end**

---

where

i	Input combination at which the output is high at least once.
$O\_Var[i]$	Number of variations in the output, for i.
$Case\_I[i]$	Number of times the input combination i occurs in the simulation data.

 The value of $Case\_I[i]$ is always equivalent to the length of its corresponding output data stream. In other words, if any input combination occurs four times, then the length of its corresponding output stream would also be four, because for every input there will be an output.

In the example shown in Fig. 5.1, the estimated fraction of variations—FOV_{EST}, for input combinations 00 and 11, are 2/1850 and 7/3050, respectively. This indicates that only a small fraction of output, in comparison to its whole size for specific input combination, is varied. This estimated fraction of variation, FOV_{EST}, is compared with the user-defined fraction of variation, FOV_{UD}, and the results are accepted if the estimated value is less than the one defined by a user. In this chapter, we assume that a user allows up to 25% variation ($FOV_{UD} = 0.25$) in the output data streams.

However, this filter alone is not sufficient to obtain the correct Boolean logic of a model. As in the case of the example shown in Fig. 5.1, the algorithm considers obtaining the output high for both input combinations 00 and 11, based on the estimated value of FOV_{EST}, and end up obtaining the XNOR logic for this circuit model. Therefore, in order to handle this situation, another filter is applied according to Eq. 5.2, which checks if the number of 1s' in the output binary data stream, for the specific input combination, are greater than half the size of the whole output data stream.

$$HIGH\_O[i] > \frac{Case\_I[i]}{2} \tag{5.2}$$

where

i	Input combination at which the output stream is being checked
$HIGH\_O[i]$	Number of 1's in the output stream corresponding to the input combination i.
$Case\_I[i]$	Number of times the input combination i occurs in the simulation data. This is equivalent to the length of corresponding output data stream as well.

For the example shown in Fig. 5.1, this condition holds false for the input combination 00 (3 $\ngtr$ 1850/2), but turns true for the input combination 11 (1875 > 3050/2). This filter also helps in making sure that the output, for a specific input combination, is certain—either high or low. Nevertheless, this filtration technique may also produce wrong results if not applied together with the first technique mentioned above. In order to understand this, consider the example case

Input Combinations (i)	Output Data Stream Case_O[i]	Number of 1s HIGH_O[i]	Case_I[i] = size of Case_O[i]	O_Var[i]
00	0101011111	7	10	5
11	0001111111	7	10	1

Input Combinations (i)	Filtering Condition 1 FOV$_{EST_i}$= O_Var[i]/Case_I[i]		Filtering Condition 2 HIGH_O[i] >Case_I[i]/2
00	= 5/10 = 0.5 (50% variable output)		7> (10/2) → TRUE
11	=1/10 = 0.1 (10% variable output)		7> (10/2) →TRUE

Fig. 5.2 Effectiveness of filtration process using both filters. An example showing how both filters are useful, when applied together, in obtaining the correct Boolean expression. © [2017] IEEE. Reprinted, with permission, from [7]

shown in Fig. 5.2, where the output binary data streams of two different input combinations, 00 and 11, are shown. The number of 1s in the output stream, for both the cases, is same; however, the patterns are different. That is, the output, for the input combination 00, remains high for the same number of times it is high for the input combination 11, but the output is highly oscillatory in the former case. The algorithm therefore discards (in this case if FOV$_{UD}$ ≤ 0.5) this unstable output and does not consider it while constructing the Boolean expression.

The pseudo code of this sub-procedure is shown in Algorithm 5.4. In order to filter out the results, both of the above mentioned conditions should be true. That is, the filtration is performed, if the estimated fraction of variation, FOV$_{EST}$, is less than the user-defined value (FOV$_{UD}$); and if the number of times the output is high for input case i is greater than half of the occurrence of input case i throughout the simulation, as shown in line 12 of Algorithm 5.4. For each filtered results, the Boolean expression is constructed through the pseudo code shown in the lines 13–26. In the end, algorithm estimates the percentage fitness of estimated Boolean expression (PFoBE), in the simulation data, according to Eq. 5.3.

$$\text{PFoBE} = 100 - \frac{\sum_i \text{FOV}_{EST_i}}{nc} \times 100 \tag{5.3}$$

where

i	Input combination at which the filtered output stream is high.
FOV$_{EST_i}$	Estimated fraction of variation in the output data stream for ith input combination.
nc	Total number of input combinations.

Algorithm 5.4: Pseudo code of the procedure **ConsBoolExp**

---

input : *O_Var, Case_I, HIGH_O, nc, N, FOV_{UD}, Case_O*
output: *BoolExpres, PFoBE*
1 **begin**
2   *Set* TV = 0; // Total variation for all input cases, nc
3   *Set* FOV_{EST} = 0; // FOV_{EST} = Estimated fraction of variation
4   *Set* Bin[*N*] = 0;
    // holds N-bit binary value of input case, nc
5   *Set* Inter_Expr = *null*;
6   *Set* Curr_Expr = *null*;
7   *Set* BoolExpres = *null*;
8   *Set* PFoBE = 0;
    // loop through all input cases nc
9   **for** *all i ∈ nc* **do**
10   **if** *(HIGH_O[i] ≥ 1)* **then**
        // Estimating fraction of variation.
11   FOV_{EST} = O_Var[i]/Case_I[i];
12   **if** *((FOV_{EST} < FOV_{UD}) AND (High[i] > Case_I[i]/2))* **then**
13   Bin[*N*] = i_d;
          // loop through all input bits
14   **for** *all j ∈ N* **do**
15   **if** *(j^{th} bit of Bin == 0)* **then**
              // put a bar (') with the name of jth input
16   Curr_Expr = In$_j$';
17   **else**
              // directly extract the name of jth input
18   Curr_Expr = In$_j$;
19   **end**
20   Inter_Expr = Inter_Expr · Curr_Expr;
21   **end**
22   **else**
23   Inter_Expr = null;
24   **end**
25   BoolExpres = BoolExpres + Inter_Expr;
26   Set Inter_Expr = null;
27   **end**
28   TV = FOV_{EST} + TV;
29   **end**
30   PFoBE = 100 - (TV/nc × 100)
31 **end**

5.2   Experimentation by Simulation

Now the algorithm is developed, it can be tested using the SBML model of any genetic logic circuit. In this book, the results of 15 genetic logic circuits models are produced. The circuits include 1, 2, and 3-inputs genetic logic circuits, which are composed of 1–7 genetic logic gates containing 3–26 genetic components. The five genetic circuit models are obtained from [3] and the remaining 10 are the models of real genetic circuits acquired from [1]. There are a total 60 circuits published in [1], which were first designed on a tool, named *Cello* [1], with the help of a hardware description language for living cells. They were then fabricated and tested in a laboratory. Out of these 60 circuits, 45 of them worked correctly in the lab. Out of those 45 functional circuits, 10 of them are chosen randomly (with different level of complexity) for testing in this book. Before discussing the logic verification results, it is important to briefly discuss the process of SBOL to SBML conversion.

5.2.1   Analysis of the SBOL-SBML Converted Genetic Circuit Models

As stated in Chap. 4 (in Sect. 4.2.6), that *Cello* generates the SBOL files. To simulate these circuits, we first need to have their SBML files. Therefore, the SBOL files of the ten circuits are first converted into SBML models using the SBOL-to-SBML conversion application in the tool named iBioSim [6]. While analyzing the SBOL/SBML models of the circuits in iBioSim [2], it was noticed that the inputs B and C of all circuits are swapped in comparison to their original circuit diagrams shown in [1]. For example, consider Fig. 5.3 in which the original circuit schematic, the SBOLv diagram, truth table, and the converted SBML model of the genetic circuit 0x0B, obtained from [1], are shown in Fig. 5.3a–d, respectively. In Fig. 5.3a and b, the inputs A, B, and C, correspond to P_{Tac}, P_{Tet}, and P_{Bad} in Fig. 5.3c and d, respectively. In Fig. 5.3b, the solid black distributions are experimental data; and blue and red line distributions are computational predictions from Cello [1], which describes the logic states 1 and 0, respectively. The logic states, for example +/+/−, indicate that the inputs C, B, and A (in order) are in logic 1/1/0 states, respectively.

As mentioned before in Sect. 4.2.6 that the SBOL file generated by Cello does not include the input sensor block for the circuit (which includes the input inducers); thus these inducers are also not included during the SBOL-SBML conversion process. Hence, the input inducers are manually added in the SBML model using iBioSim, as shown in Fig. 5.3d with the red-dotted block. The rest of the model (shown as blue-dotted block) is the result of the SBOL-SBML conversion process [6]. The external inputs, in all these ten circuits, are the inducers IPTG, aTc, and Arabinose, which control the activities of the promoters P_{Tac}, P_{Tet}, and P_{Bad}, respectively, as shown in Fig. 5.3d.

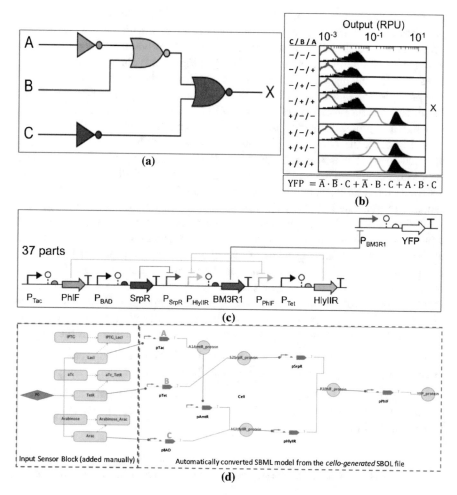

Fig. 5.3 Genetic circuit 0x0B [1]. (**a**) Circuit schematic (**b**) Truth table and (**c**) SBOLv diagram [Image courtesy of [1]]. (**d**) Screen shot of the auto-generated SBML model in [2] using SBOL-SBML converter [6]

It is also important to mention that, in Cello [1], a circuit is generated by a random search of compatible genetic gates using the simulated annealing algorithm [5]. Since the circuit is created using the non-deterministic search, the solution may be different every time the process is executed. This is why the genetic components shown in Fig. 5.3c are different from those depicted in Fig. 5.3d. Despite having different genetic components, the circuit structure should not be changed in order to achieve the same functionality of each circuit. In Fig. 5.3a, the input A (or P_{Tac}) is connected to a NOT gate that produces the *PhlF* protein, which in turn suppresses the output promoter P_{PhlF}, as shown in Fig. 5.3c. The input B (or P_{Tet}) is connected directly to one of the inputs of the NOR gate (producing *HlYllR* protein), as depicted in Fig. 5.3a and c. Another input of this NOR gate is the output promoter of the first NOT gate (i.e. P_{PhlF}), which together with the input B (or P_{Tet}) generates the protein $HlYllR$ and suppresses the output promoter, P_{HlYllR}, as shown in Fig. 5.3c. Similarly, the input C (or P_{Bad}) in Fig. 5.3a is connected to a separate NOT gate (producing $SrpR$ protein) which suppresses the output promoter P_{SrpR}. The promoter P_{SrpR} together with the output promoter P_{HlYllR} of the first NOR gate (producing $HlYllR$ protein) performs the NOR logic to produce the protein $BM3R1$, which in turn produces the final output, yellow fluorescent protein (YFP).

In Fig. 5.3d, the NOT gate of input A is producing the protein $A1\_AmtR$, as opposed to what it is shown to produce ($PhlF$ protein) in Fig. 5.3a and c. This is because of the non-deterministic gates assignment in Cello, which may have different output proteins, but the functionality (the NOT logic in this case) remains same. The generated protein, $A1\_AmtR$, suppresses the promoter, P_{AmtR}, thus exhibiting the NOT logic. However, it can be noticed, in Fig. 5.3d, that instead of input B (or P_{Tet}), the input C (or P_{Bad}) is directly connected to one of the inputs of the NOR gate (producing $HlYllR$ protein) along with the promoter P_{AmtR}. The input B (or P_{Tet}) in Fig. 5.3d, which should be connected directly to a NOR gate, seems to be routed to a NOT gate (producing $SrpR$ protein). In other words, the inputs B and C of the original circuit 0x0B are swapped. Due to this problem of swapped B and C inputs, the functionality of any genetic circuit shown in [1] would be changed. For example, the optimized Boolean expression of the original circuit 0x0B is $(\overline{C} + \overline{\overline{A} + B})$, in which there is a NOT gate with input C and an intermediate NOR gate with inputs $\overline{A}$ and B. If the inputs B and C are swapped, then the expression will become $(\overline{B} + \overline{\overline{A} + C})$, having NOT gate with input B and an intermediate NOR gate with inputs $\overline{A}$ and C, which clearly changes the functionality of the circuit.

Be careful while analyzing the genetic circuit models presented in [1]. The inputs B and C of all the circuits are swapped during their conversion from SBOL to SBML.

5.2.2 Logic Analysis and Verification

The SBML files generated for ten circuits [1] (with inputs B and C swapped) and for five other circuits [3] are used in D-VASim to perform the experimentation followed by the logic verification. The external inputs in the circuits obtained from [1] are IPTG, aTc, and Arabinose, which were varied in the experimentation to observe their logical behaviors.

Let us examine the results of the genetic circuit, 0x0B, which is run for 10,000 simulation time units. Also, assume that the value of propagation delay is 1000 time units. Furthermore, the upper and lower threshold values of the circuit is equal to 15 and 0 molecules, respectively.

Each input combination must be applied for the amount of time equivalent to the value of propagation delay. In this experiment, each input combination must be held at least for 1000 time units.

The screen shot of the interactive experimentation of genetic circuit, 0x0B, is shown in Fig. 5.4. The upper half of this figure indicates the interactive analog simulation results in which the concentrations of external inputs, *IPTG, aTc*, and *Arabinose* are varied to observe the behavior of the output protein, *YFP*. It is evident from this figure that it is not easy to grasp the logic of a genetic circuit model with these messy analog waveforms. On the other hand, the Boolean logic is a bit easier to analyze in the corresponding digital waveforms shown in Fig. 5.4. The logic verification algorithm made this analysis further easier by automatically analyzing the whole simulation data (shown in Fig. 5.4) and indicating the results in the form of a raw Boolean expression as shown in Fig. 5.5.

The percentage fitness of the Boolean expression in the simulation data is also shown in Fig. 5.5. The logic verification algorithm expresses the logical behavior in the *Sum of Products (SoP)* form of Boolean expression (See Chap. 6 for more details). The original Boolean expression of the circuit 0x0B, obtained from the truth table shown in Fig. 5.3b, is $\overline{A} \cdot \overline{B} \cdot C + \overline{A} \cdot B \cdot C + A \cdot B \cdot C$. However, the Boolean expression for the same circuit, obtained in these experimentation, is shown

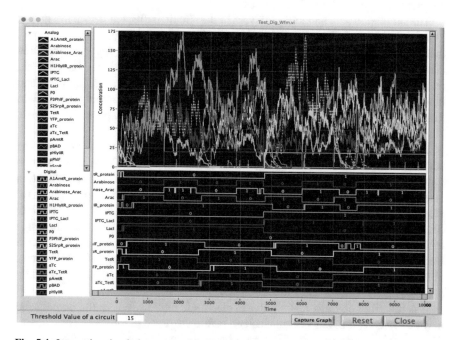

Fig. 5.4 Interactive simulation traces of 0x0B, with its corresponding digital waveforms, for logic analysis

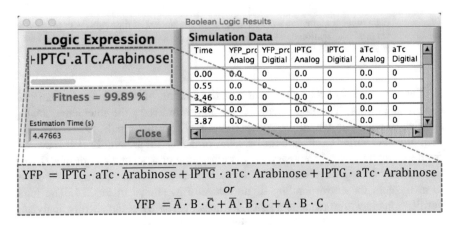

Fig. 5.5 Logic analysis results of the genetic circuit 0x0B in D-VASim

in Fig. 5.5, which is equivalent to $\overline{A} \cdot B \cdot \overline{C} + \overline{A} \cdot B \cdot C + A \cdot B \cdot C$; where inputs A, B, and C correspond to IPTG, aTc, and Arabinose, respectively. This is because the inputs B and C were swapped in the SBML model (see Sect. 5.2.1). This indicates that the proposed logic analysis algorithm estimates the correct logic of the SBML model being tested, provided that the experimentation is performed with the correct propagation delay and threshold values.

The simulation data and logic analyses of the circuit, 0x0B, are shown in Fig. 5.6. The results shown in Fig. 5.6 are used to obtain the logical behavior of the circuit 0x0B. In Fig. 5.6, *Case_I* indicates the number of times each input combination occurs during the total 10,000 time units of simulation. It further includes the number of times the output of a circuit remains high, *High_O*, for that particular input combination along with the number of variations in the output data, *O_Var*. Also, the input combinations, at which the circuit's output is expected to be high, are highlighted in green color along the x-axis.

In Fig. 5.6, the output variation of circuit 0x0B is not too high. For example, the output state appears to be logic-1 for the input combination 100 and seems quite stable having very low variation value of 2. The reason why the input combination 100 has so many logic-1 output states is due to the fact that the output is high for the previous input combination 011. When the input combination is changed from 011 to 100, the output starts to decay gradually, and remains high until it passes by the threshold level. This input combination should, therefore, be included in the Boolean expression, but however filtered out using Eq. 5.2, because for 3587 times of the input combination 100 occurs during the entire simulation, the corresponding output remains high for 1191 times (<3587/2).

It is therefore obvious that similar to electronic circuits, where the output state may be incorrect if the inputs are changed before the propagation delay has elapsed, the correct behavior of a genetic circuit can only be obtained when each possible input combination is applied for sufficient amount of time.

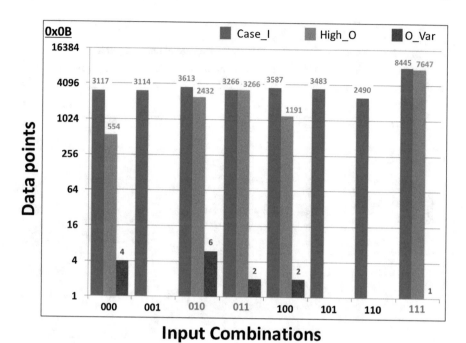

Fig. 5.6 Analytical simulation data of the genetic circuit model 0x0B [1]. © [2017] IEEE. Reprinted, with permission, from [7]

5.2.3 Effects of Varying Threshold Value on the Behavior of Genetic Circuit

The logical behavior of genetic circuits is also analyzed by varying the threshold value of input concentrations to very low (3 molecules) and very high (40 molecules), and observed that the same circuits behave differently.

Figure 5.7 shows the comparison of simulation data for the circuit 0x0B for the above mentioned two threshold values. In this figure, it can be noticed that the output response of 0x0B circuit, for upper threshold value of three molecules, is entirely different and it behaves like a 3-input AND gate. This is because the applied input concentration is too weak to trigger the output concentration; but when applied together i.e., 111, the output is triggered high to satisfy the applied filters. On the other hand, 0x0B circuit has two wrong states (shown in the Boolean expression in Fig. 5.7) when 40 molecules are applied as an input concentration. For this case of threshold value, the output response also seems to oscillate between logic high and low for large number of times (Fig. 5.7) as compared to when its threshold value is set to 15 (Fig. 5.6). This is because the concentration levels of input and output species are not clearly distinguishable when the applied input concentration is high.

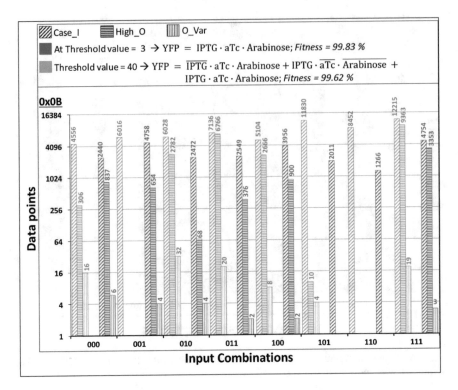

Fig. 5.7 Analytical data of circuit 0x0B for threshold values 3 and 40. © [2017] IEEE. Reprinted, with permission, from [7]

5.2.4 Performance Analysis

The performances of the proposed algorithm, on all 15 circuits, are also analyzed, as shown in Fig. 5.8. Depending upon the complexity of a circuit, each circuit may have different amount of data for specified simulation time.

For example, there is only one gate, composed of three genetic components, in genetic NOT gate circuit. For this small circuit, only 3.5×10^3 reactions occurred during 10,000 simulation time units. In contrast, a bigger genetic circuit (x1C) containing 7 cascaded gates, composed of 26 genetic components, had 43.1×10^3 reactions executed during the same simulation time. Therefore, the size of simulation data for NOT gate circuit is much smaller than the size of x1C circuit.

Due to this, the time to estimate the logic of a circuit is much lower for a NOT gate circuit as compared to the time required to estimate the logical behavior of the x1C circuit. This is shown as *data points* in Fig. 5.8, which corresponds to the number of reactions executed during 10,000 simulation time units. Similarly, the difference of analysis time for the circuits having same number of gates reflects the

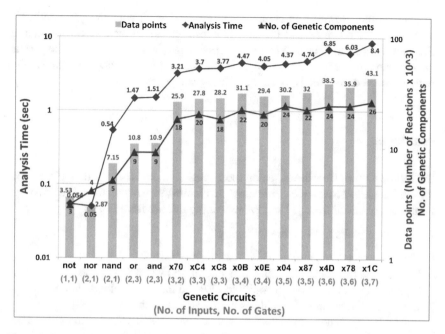

Fig. 5.8 Performance evaluation of the proposed algorithm over 15 circuits. First five circuits are from [3] and the remaining ten circuits are from [1]

difference in the number of genetic components, and thus in the number of *data points*.

5.3 Summary

In this chapter, a methodology to analyze and verify the logical behavior of a genetic circuit is presented. It is shown through simulation experiments that the circuit may not behave as intended if the circuit parameter(s), like threshold value, are varied. This may help users to analyze the circuit's behavior and robustness for different parameter sets before creating them in the laboratory. Furthermore, the performance of the proposed algorithm is analyzed over the number of genetic circuit models, and observed that it takes about 8.4 s to analyze the logic of a complex genetic circuit with significantly large-sized data. As the experimentation in the laboratory requires a couple of hours to analyze even a single output state [1], the discussed simulation-based approach is likely to be useful for genetic circuit designers to analyze the intended logic of genetic circuits prior to their implementation and testing in the laboratory. The proposed algorithm is scalable and can be used to analyze genetic circuit with any number of inputs. However, when the circuit inputs are increased, the size of simulation data will also be increased, which results in the rise of logic estimation time. However, in comparison to electronic circuits, the genetic circuits are much more complex and difficult to construct due to their stochastic nature. This fact suggests that the size of genetic circuits may not grow with a same pace as the size and complexity of electronic circuits were increased, and thus the studied method can be used effectively for genetic logic analysis.

Problems

The following problems are based on the 14 SBML models of genetic logic circuits enclosed in http://Resources/Problems/Chapter5/Models/ directory. The first ten circuits are obtained from [1] and the remaining five are from [3].

5.1 For the circuits 6–14 do the following:

1. Check the electronic circuit schematic of a model in [1].
2. Looking at the RPU chart of each circuit shown in [1], write the Boolean expression in terms of output variable *YFP* and input variables *A, B,* and *C*.
3. Check the corresponding SBOL diagram of a circuit in [1].
4. Load the SBML model in iBioSim and compare the SBOL diagram of a circuit shown in [1] with the one loaded in iBioSim. Verify if the inputs *B* and *C* are swapped.

iBioSim can be downloaded from https://github.com/MyersResearchGroup/iBioSim/releases.

5.2 For all the 14 circuits, assume the following parametric values:

- Threshold Value = 15.
- Propagation delay = 1000.

Now perform interactive stochastic experimentation for 10,000 time units; apply all the possible input combinations, and do the following:

1. Open the mixed signal waveform window and observe the digital waveforms in order to infer the Boolean logic function of the model.
2. Close the mixed signal waveform viewer, and execute the automated **Logic Verification**. Select YFP in the **Select Output Specie** drop-down option and specify the above mentioned threshold value for all species.

To view mixed signal waveforms, drop the **Options** menu down from the toolbar and select **Mixed Signal Waveforms**. Read DVASim QSG for more details.

The **Mixed Signal Waveforms** window can only be seen after running the simulation.

5.3 The source code of DVASim is publicly available and hosted at the GitHub repository https://github.com/hasanbaig/D-VASim. Clone DVASim source code on your local machine and execute the following steps to make the observations required in this exercise.

LabVIEW 2019 (or later) must be installed on your machine to perform this exercise.

Going through Sect. 7.3.3 may help in understanding the following steps.

1. Open the file **BLE v2.0-scalable.vi** placed under https:/SubVIs/ directory.
2. Maximize the front panel, where you will see three additional indicators—*Case_I, HIGH_O*, and *O_Var*.
3. Now open the block diagram where you will see a **Diagram Disable Structure** and a **Stacked Sequence Structure**.
4. Enable the structure labelled **"A"** in the **Diagram Disable Structure** placed outside the **Stacked Sequence Structure**.
5. Go to sequence 2 in the **Stacked Sequence Structure** and enable the structure labelled **"X"** in the **Diagram Disable Structure** placed inside the **Flat Sequence Structure**.

Once done with the steps mentioned above, go to the front panel and select the simulation data file *"not_RB_SSADataLog.xls"* placed under http://Chapter5/Solutions/5.3/1.not/ directory. Run the VI; select the appropriate output species when prompted; and press OK. Note down the values of *Case_I, HIGH_O*, and *O_Var* and observe the stats of interactive experimentation.

These values are listed in successive order of input combinations. For example, for two-input circuits, the order of input combinations are 00, 01, 10, 11.

To stop executing this VI, go to block diagram and press the **Abort Execution** button. Do not press the **Close** button on the front panel, otherwise it will close the VI.

Now select the simulation data of all other models one by one and observe the stats for each of them. Produce the results similar to the one shown in Fig. 5.6.

On the block diagram, you have to delete and reconnect the wire with specific nodes for different model's data. Check the comments in the outer **Diagram Disable Structure** on the block diagram.

Make sure to revert back the steps 4 and 5 mentioned above.

References

1. A.A.K. Nielsen, B.S. Der, J. Shin, P. Vaidyanathan, V. Paralanov, E.A. Strychalski, D. Ross, D. Densmore, C.A. Voigt, Genetic circuit design automation. Science **352**(6281), aac7341–aac7341 (2016)
2. C.J. Myers, N. Barker, K. Jones, H. Kuwahara, C. Madsen, N.P.D. Nguyen, iBioSim: a tool for the analysis and design of genetic circuits. Bioinformatics **25**(21), 2848–2849 (2009)
3. C.J. Myers, *Engineering Genetic Circuits* (Chapman and Hall/CRC, New York, 2009)
4. H. Baig, J. Madsen, Simulation approach for timing analysis of genetic logic circuits. ACS Synth. Biol. **2**, acssynbio.6b00296 (2017)
5. D.T. Gillespie, A general method for numerically simulating the stochastic time evolution of coupled chemical reactions. J. Comput. Phys. **22**(4), 403–434 (1976)
6. N. Roehner, Z. Zhang, T. Nguyen, C.J. Myers, Generating systems biology markup language models from the synthetic biology open language. ACS Synth. Biol. **4**(8), 873–879 (2014)
7. H. Baig, J. Madsen, Logic analysis and verification of n-input genetic logic circuits, in *Design Automation and Test in Europe (DATE) 2017 (Lausanne)* (IEEE, New York, 2017), pp. 654–657

Chapter 6
Technology Mapping of Genetic Circuits

In previous chapters, we have talked about simulation and analysis of genetic logic circuits using the SBML models. We also mentioned earlier that the SBML standard is used to represent the behavior of a genetic circuit using reaction kinetics, whereas the SBOL standard is used to represent the structure of a circuit and describes how the intermediate genetic components are connected together.

In contrast to electronic logic gates, which have the same physical quantity, i.e., voltage, at the input and output, the genetic logic gates have different quantities acting as an input and output. This makes it very challenging to integrate genetic logic gates to construct complex genetic circuits because the triggering molecules relating input and output, between cascaded gates, have to be compatible and unique.

In this chapter, a new standalone tool, *GeneTech* (name extracted from *Genetic Technology* mapping), is discussed, which automate the process of generating genetic circuits for dedicated Boolean functions. *GeneTech* performs Boolean optimization, followed by synthesis and technology mapping using a library of genetic logic gates. It uses the gates library which is developed and tested in the laboratory by MIT and Boston University [2]. *GeneTech* takes the Boolean expression of a genetic circuit as input, and optimizes it first. Then, it synthesizes the optimized Boolean expression into NOR-NOT form in order to construct the circuit using the real NOR/NOT gates available in the genetic gates library [2]. In the end, *GeneTech* performs technology mapping to generate all the feasible circuits, with different genetic gates, to achieve the desired logical behavior. The circuits are generated in the form of standard SBOL notation.

Electronic Supplementary Material The online version of this chapter (https://doi.org/10.1007/978-3-030-52355-8_6) contains supplementary material, which is available to authorized users.

81

6.1   Introduction and Motivation

The architecture of GeneTech is originally inspired from the processes of optimization and technology mapping of electronic circuits in the EDA industry. In EDA
for digital electronics, the combinatorial circuit optimization is always required
to implement the circuit with the minimum number of logic gates [4]. This area-
efficient implementation of digital circuits not only helps reducing the size of
electronic devices but also avoids wasting power and redundant resources.

   In order to get the insight of logic optimization, consider the digital circuit for
the expression $ab + b + ac$, in Fig. 6.1a. This figure shows that the original circuit
contains four logic gates. After running the optimization algorithm, the number of
gates in the circuit reduces down to two while preserving the actual functionality, as
illustrated in Fig. 6.1b.

   The 2-input genetic NAND gate is termed universal in [3] because it is possible
to construct other combinational gates by cascading the collection of them. For
instance, genetic inverter and a 2-input genetic NAND gate can be used to construct
a genetic XOR gate. Suppose L and T represent the two genetic inputs, LacI and
TetR, then the function of a genetic XOR gate can be described by

$$L \oplus T = \overline{L}T + L\overline{T} \tag{6.1}$$

   The standard schematic of a XOR gate is shown in Fig. 6.2. It consists of AND,
OR, and NOT gates. The direct genetic implementation of AND and OR gates are
not shown in [3]. However, as mentioned above, these gates can be constructed with
the help of universal 2-input genetic NAND gates and genetic inverters [3].

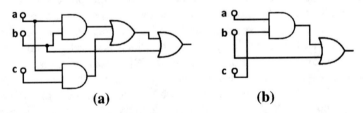

Fig. 6.1 Digital circuit of the expression $ab + b + ac$. (**a**) Original circuit containing four gates.
(**b**) Optimized circuit having two gates

Fig. 6.2 Standard schematic
of the XOR gate

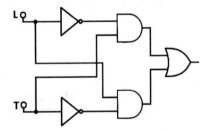

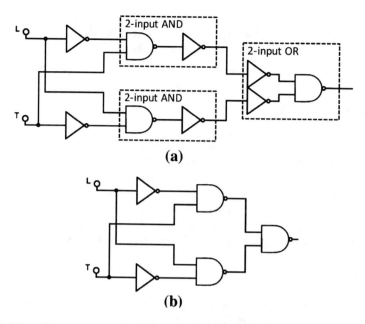

Fig. 6.3 Schematic of the genetic XOR gate (**a**) constructed with 2-input NAND gates. (**b**) The optimized circuit

The equivalent circuit of a XOR gate constructed with the available genetic components is shown in Fig. 6.3a. This schematic of a genetic XOR gate needs to be optimized with the following two constraints—the functionality of a circuit should remain same, and the components of the optimized circuit must be available in the library of genetic gates. It can be noticed that the back-to-back inverters are present in Fig. 6.3a resulting in the following equivalent Boolean expression.

$$ L \oplus T = \overline{\overline{\overline{LT}} \cdot \overline{L\overline{T}}} \tag{6.2} $$

One of the methods to estimate the cost of a circuit is to calculate the number of inputs to each gate—hence the cost is higher if the number of gates is increased. For the circuit shown in Fig. 6.3a, the cost of three NAND gates is 6 and six NOT gates is also 6, so the total cost is 12, which can be reduced down to 8 as shown in Fig. 6.3b. This optimization seems straightforward and acceptable as it fulfills the above mentioned constraints and is composed of inverters and 2-input NAND gates, which are available in the genetic gates library [3]. However, the optimization and technology mapping of genetic circuits is not similar to electronic circuits. This is because the input and output quantities of electronic circuits are the same, i.e., voltage, and therefore the electronic gates can easily be cascaded together. On the contrary, the input and output quantities of genetic gates are different, and therefore the signal matching has to be considered while mapping genetic gates on the circuit.

Similar to the digital electronic circuits, we want to avoid having redundant logic in genetic circuits. Therefore, the logic expression, either for digital or genetic circuits, needs to be minimized using any optimization technique.

6.2   Methodology

A digital circuit or Boolean logic can be expressed either in the minterm or maxterm canonical form. Minterms are called products because they are the logical AND of a set of variables/literals and maxterms are termed as sums because they are the logical OR of a set of variables/literals. Therefore, the Boolean expression can either be expressed as sum of minterms/products (SOP) or product of maxterms/sums (POS). In the example shown in Eq. (6.3), the left-hand side represents the SOP form and the right-hand side represents its equivalent POS form.

$$ab + b + ac = (a + b + c)(a + b + \overline{c})(\overline{a} + b + c) \qquad (6.3)$$

GeneTech is able to process the Boolean expressions available in both SOP and POS forms. The design flow of *GeneTech* is shown in Fig. 6.4. It takes the Boolean expression and first optimizes it followed by the synthesis and technology mapping. For each possible solution, GeneTech generates the output in three different forms— SBOL data file, SBOL Visual notation, and equivalent electronic circuit schematic. Each of the highlighted steps shown in Fig. 6.4 is described separately in the following subsections.

6.2.1   Logic Optimization

The digital logic minimization using meta-heuristics is a classical optimization problem and has already been addressed before [5]. As demonstrated above, the minimization of genetic logic circuits follows the same procedure as that of digital circuits. Therefore, meta-heuristics can also be employed to optimize genetic logic circuits. The algorithm used in *GeneTech* for Boolean expression optimization is based on the simulated annealing algorithm [6].

In order to apply simulated annealing to any optimization problem, one must define the search space, the neighbor selection method, acceptance probability, and the annealing schedule. The initial temperature, the rate at which the temperature reduces, the number of iterations at each temperature, and the stopping criterion are known as the annealing schedule [5].

Suppose, **B**, be a search space, is a set of all possible Boolean expressions implementing a Boolean function $f : \{0, 1\}^n \rightarrow \{0, 1\}^m$. During each iteration, the annealing algorithm considers some neighboring state expression, $E_N \in \mathbf{B}$ of the current state expression $E_C \in \mathbf{B}$, and decides probabilistically either to move to

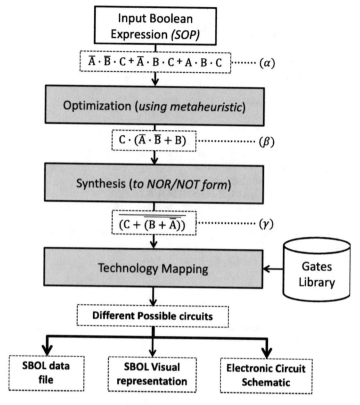

Fig. 6.4 The technology mapping flow of *GeneTech*

state E_N or remain in the state E_C. The pseudo code of annealing-based algorithm for Boolean expression minimization is shown in the Algorithm 6.1. The objective function for the minimization of the Boolean expression is

$$Minimize \sum L$$

where L is the total number of literals in the expression.

That is, the objective is to minimize the cost of a Boolean expression in terms of literals. This minimization is carried out by the help of Boolean replacement rules listed in Table 6.1. The algorithm starts by taking the inputs—Boolean expression, temperature coefficient, initial temperature, and a time to execute the algorithm. The procedure *COST_CAL* calculates the initial cost of the Boolean expression (line 2). The while loop runs until the end time, t_E, lapsed (line 5). The procedure, *SEARCH-NEIGHBOR* (line 6), obtains the neighboring state expression, E_N, by applying the Boolean replacement rules listed in Table 6.1.

Algorithm 6.1: Pseudo code of *logic optimization*

---

    input : E_0, T_{COF}, T_0, t_E
    output: E_B, C_B
    `// `E_0` = Initial expression, `T_{COF}` = Temperature coefficient`
    `// `T_0` = Initial temperature, `t_E` = End time to finish`
    `   algorithm,`
    `// `E_B` = Best expression, `C_B` = Best cost`

1 **begin**
2 $C_0 = \textbf{COST\_CAL}(E_0);$ `// `C_0` = Initial cost`
3 $E_C = E_0;$ `// `E_C` = Current expression`
4 $T_C = T_0;$ `// `T_C` = Current temperature`
5   **while** *($t_C < t_e$)* **do**
        `// `t_C` = Current time`
6 $(E_N, C_P, C_N) = \textbf{SEARCH-NEIGHBOUR }(E_C);$
        `// `E_N` = Neighbouring expression, `C_P` = Previous cost,`
        `     `C_N` = New cost`
        `// * Calculate acceptance probability (AP) *`
7   **if** *($C_N > C_P$)* **then**
8 $AP = \textbf{CALC}(T_C, C_P, C_N);$
9   **else**
10 $AP = 1;$
11   **end**
        `// * Update new expression *`
12   **if** *($AP > random[0,1)$)* **then**
13 $E_C = E_N;$
14   **else**
15 $E_C = E_C;$
16   **end**
17 $T_C = T_{COF} \times T_C;$
18   **end**
19 $E_B = E_C;$
20 $C_B = C_N;$
21 **end**

Table 6.1 Boolean replacement rules

Case	Replacement
x + x (x.1) or (1.x) (x + 0) or (0 + x)	x
x' + x' (x'.1) or (1.x') (x' + 0) or (0 + x')	x'
(x + 1) or (1 + x) (x' + 1) or (1 + x') (x + x') or (x' + x)	1
(x.0) or (0.x) (x.x') or (x'.x)	0
x.y + x.z	x.(y + z)

The annealing algorithm calls the procedure, *SEARCH-NEIGHBOR*, by passing the current state of expression as an input. It then randomly decides whether to expand any of the minterm in the expression or proceeds without expanding it. The re-expansion of already reduced minterms is sometimes useful to avoid getting stuck in local optimum value. The total number of minterms in the Boolean expression is then calculated and a combination of any two minterms is selected randomly for minimization. The number of literals in each minterm is then evaluated. If both of the minterms contain single literal only, they directly scan through the replacement rules followed by the construction of a new expression. If the number of literals, in any of them, is greater than one, then the algorithm performs a check if a common literal(s) is present in both of them. If no match is found, the input expression is reconstructed. If a common literal is found, the algorithm arranges the minterms in the form

$$M(R_{L-1} + R_{L-2})$$

where
M is the matched literal(s).
R_{L-1} corresponds to rest of the literals in the first minterm.
R_{L-2} corresponds to rest of the literals in the second minterm.
    The procedure, SEARCH-NEIGHBOR, then checks the nested elements (elements inside braces) of the reduced expression for further reduction, and passes them through the Boolean replacement rules. The process then searches for the nested minterms containing braces and keeps searching until all the nested minterms, containing braces, are passed through the Boolean replacement rules. Finally, all the possible combinations, within nested expression, are checked for common literals. Unlike the beginning of the algorithm, where the combination of two minterms is randomly chosen, all the combinations are checked one by one at this stage. The new expression, E_N, is then constructed and a new cost of a reduced expression, C_N, is calculated. As an example, consider the same SOP form of the expression (6.3), which is rewritten as expression (6.4) below

$$ab + b + ac \tag{6.4}$$

To minimize the expression (6.4), the algorithm randomly chooses the combination of, say, first and third minterms initially, i.e., ab and ac. As literal a is common in both of them, the expression reduces down to the expression (6.5).

$$a(b + c) + b \tag{6.5}$$

Since these two minterms are selected randomly, it is possible that this combination may not produce the optimized result. Because of this uncertainty, the capability of expanding the reduced minterms is incorporated in the algorithm. The reduction of expression (6.4)–(6.5) results in the reduction of cost from 5 to 4 literals with no further reduction possible. The circuit generated for this expression is shown

Fig. 6.5 The local optimized
circuit of an example
expression (6.4)

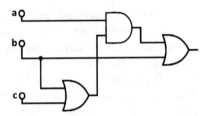

in Fig. 6.5, which is not the optimized one. With the capability of expanding the
reduced minterms, the algorithm is able to come out of this local optimized solution.
Therefore, in the above example, if the expression (6.5) is expanded again into (6.4)
and a combination of first and second elements is chosen, i.e., ab and b, then the
optimized cost of the expression (6.4) would reduce from 5 literals to 3 literals, as
shown below, and would result in the circuit pictured in Fig. 6.1b.

$$b(a + 1) + ac \tag{6.6}$$

$$b + ac \tag{6.7}$$

Finally, the procedure *CALC* calculates the acceptance probability based on the
cost of Boolean expressions (lines 7–11). If the new cost, C_N, is less than the
previous cost, C_P, the new solution is assigned a probability equals to one and
accepts the new expression as a current best solution. Otherwise, the acceptance
probability of new expression is calculated according to the following expression:

$$AP = e^{\frac{-(C_N - C_P)}{T_C}} \tag{6.8}$$

Then a random number between 0 and 1 (exclusive) is generated. The algorithm
accepts the new expression as a best solution if the acceptance probability, AP, is
greater than the generated random number. Else it is rejected and the previous state
of expression is considered as a current best solution (lines 12–16).

6.2.2 Logic Synthesis

GeneTech is developed to be used for constructing real genetic circuits from the
genetic gates library tested in the laboratory [2]. The circuits experimented in [2] are
composed of genetic NOR and genetic NOT gates. Therefore, in this stage of logic
synthesis, the AND/NAND terms present in the optimized Boolean expression are
converted to NOT/NOR form by applying the following De Morgan's Laws.

$$\overline{AB} = \overline{A} + \overline{B} \tag{6.9a}$$

Algorithm 6.2: Pseudo code of *logic synthesis*

---

input : Exp_{in}
output: Exp_{out}
// Exp_{in} = Input expression, Exp_{out} = Output Expression
1 **begin**
2   **if** *(Exp_{in} contain braces terms)* **then**
3   **for** *(all braces terms in Exp_{in})* **do**
4   **for** *(all minterms inside braces)* **do**
5   MTIB = **ProcessANDTerms**;
                // MTIB = Array holding minterms inside braces
6   **if** *(MTIB > 1)* **then**
7   OutputString = **ProcessORTerms**;
8   OutputString = **ProcessNANDTerms**;
9   OutputString = **MintermsToExpression**;
10   **else**
11   OutputString = **MintermsToExpression**;
12   **end**
13   **end**
14   **end**
15   **else**
16   **for** *(all minterms in Exp_{in})* **do**
17   Minterms = **ProcessANDTerms**;
18   OutputString = **MintermsToExpression**;
19   **end**
20   **for** *(all minterms in OutputString)* **do**
21   **ConvertToNOR**;
22   **end**
23   **end**
24   Exp_{out} = OutputString;
25 **end**

$$AB = \overline{\overline{A} + \overline{B}} \qquad (6.9b)$$

$$\overline{A.B} = \overline{A} + B \qquad (6.10)$$

The pseudo code of the logic synthesis algorithm is shown in the Algorithm 6.2. The algorithm is briefly explained with the help of example shown in Fig. 6.6, where the input expression is shown in red (line 0) and the final output expression is highlighted in green (line 8). The algorithm begins by taking the optimized Boolean expression as an input and checking first if it contains any minterms with braces (line 2 in Algorithm 6.2). All minterms inside each braced-minterm are first passed through the procedure, *ProcessANDTerms*, to convert ANDed minterms to NOR terms (lines 4, 5 in Algorithm 6.2). These steps are shown in Fig. 6.6 in line 0 and 1. The algorithm then processes OR terms, if the number of minterms inside braces is greater than one (lines 6–9 in Algorithm 6.2). This is shown in line 2 in Fig. 6.6, where the OR terms are converted to NAND by the procedure ProcessORTerms, using Eq. (6.9a). The same procedure also expands braces by multiplying minterms

Fig. 6.6 Example expression
to illustrate how synthesis
algorithm works

0	$\bar{c} \cdot (ab + \bar{b})$	
1	$\bar{c} \cdot (\overline{a + \bar{b}} + \bar{b})$	$\because (6.9b)$
2	$\bar{c} \cdot \overline{((\bar{a} + b) \cdot b)}$	$\because (6.9a)$
3	$\bar{c} \cdot \overline{(\bar{a}b + \bar{b}b)}$	$\because$ braces multiplication
4	$\bar{c} \cdot \overline{(\bar{a}b)}$	$\because \bar{b}b = 0$
5	$\bar{c} \cdot (\bar{\bar{a}} + \bar{b})$	$\because (6.9a)$
6	$\bar{c} \cdot (a + \bar{b})$	$\because \bar{\bar{a}} = a$
7	$\bar{c} \cdot \overline{\overline{(a + \bar{b})}}$	$\because \bar{\bar{a}} = a$
8	$\overline{(c + \overline{(a + \bar{b}))}}$	$\because (6.10)$

and checks if any Boolean replacement rules (shown in Table 6.1) are applicable. This is shown in lines 3 and 4 in Fig. 6.6.

After applying Eq. (6.9a) (in line 2 in Fig. 6.6) to convert the expression from OR to NAND, it is converted back again to NAND form only if the expression is manipulated by applying the Boolean replacement rules. This NAND to OR conversion is shown in the lines 4–6 in Fig. 6.6, using the procedure, *Process-NANDTerms*, (line 8 in Algorithm 6.2). Then all of the minterms arranged in an array are converted to string expressions using the procedure, *MintermsToExpression* (line 9 in Algorithm 6.2). If there are no minterms in the expression which contain braces (e.g., $a + bc$), the algorithm then processes the ANDed minterms in the expression (lines 16–19 in Algorithm 6.2). After processing all minterms with or without braces, the algorithm then converts all minterms in the expression to NOR (lines 21–22 in Algorithm 6.2). In the example shown in Fig. 6.6, there is only one minterm in the expression with braces, which contain two sub minterms inside braces. In the last step, Eq. 6.10 converts the ANDed minterms (with braces) to NOR, as shown in the lines 7 and 8 in Fig. 6.6.

6.2.3 Genetic Technology Mapping

When the Boolean expression is synthesized into NOR/NOT form, genetic technology mapping can be performed using the genetic gates library. The genetic gates have been extracted from [2] by analyzing the SBOLv diagrams of all the circuits shown in [2] and arranging them in the separate lists of genetic NOT and NOR gates. The list of genetic NOT and NOR gates is shown in Fig. 6.7. The list shown in this figure consists of 35 NOR and 17 NOT gates, which indicates that there are

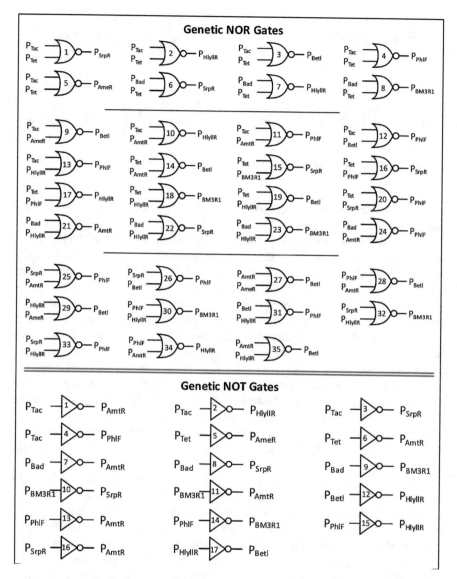

Fig. 6.7 Genetic gates library constructed from the circuits shown in [2]

several different genetic components which can be used to perform the logical NOR or NOT operations inside a living cell.

Figure 6.8 shows the equivalent SBOLv diagrams [1] of the genetic NOT-1 ("*1*" *refers to the identity of this gate shown in* Fig. 6.7) and NOR-1 gates. For example, the input of a genetic NOT-1 gate, shown in Fig. 6.8, is P_{Tac} promoter which produces the protein AmtR based on the presence or absence of isopropyl thiogalactoside (IPTG) inducer. If IPTG is absent, the input promoter P_{Tac} is active

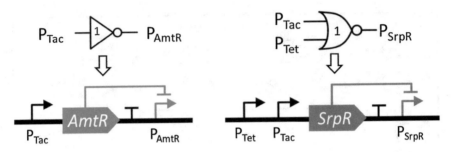

Fig. 6.8 SBOLv diagrams of NOT-1 and NOR-1 gates

(logic-1) and produces the AmtR protein, which then suppresses the activity of the output promoter PAmtR (logic-0), and vice versa.

Similarly, for the NOR-1 gate, shown in Fig. 6.8, the activities of the input promoters, P_{Tac} and P_{Tet}, are controlled by the inducers IPTG and anhydrotetracycline (aTc), respectively. When both of them are present, the activities of input promoters, P_{Tac} and P_{Tet}, are suppressed (logic-00), which reduces down the production of protein SrpR. The output promoter, P_{SrpR}, becomes active (logic-1) when the protein SrpR is not produced in a significant amount to suppress it. When either or both of these inducers are absent, the corresponding input promoter(s) becomes active (logic-01, 10, 11) and produces the protein, SrpR, which in turn suppresses (logic-0) the activity of output promoter, P_{SrpR}, thus exhibiting the NOR logic.

As discussed above, the input and output quantities in genetic gates are different. This makes it challenging to construct a genetic circuit by making sure that the output of the first gate is compatible with the input of the following one. The *GeneTech* mapping algorithm constructs a genetic circuit by using the genetic gates, from the gates library (Fig. 6.7), whose inputs and output proteins are matched with each other.

The mapping algorithm is explained briefly with the help of example shown in Fig. 6.9. In this figure, the output expression from the previous stage, *logic synthesis*, is considered to be an input to the algorithm for technology mapping. Inputs a, b, and c correspond to the external inputs, P_{Tac}, P_{Tet}, and P_{Bad}, respectively, in [2]. The algorithm works on the depth-first search approach and hence begins by mapping genetic gates first on the deepest element(s) in the expression. Therefore, the list of all the possible NOT gates for $\bar{b}$ is extracted from the genetic gates library first.

For understanding, let us name the list of inverters for $\bar{b}$ as *list-A*, which consists of NOT-5 and NOT-6 gates. The algorithm selects NOT-5 first from list-A, and then checks if its output is compatible with the input a (or P_{Tac}) in the form of NOR gate. At this step, a new list, say *list-B*, is created which contains three NOR gates with P_{Tac} as one of their inputs. Since the output of NOT-5 is compatible with the input of the NOR-9, the algorithm proceeds further with the search of NOR gate(s) with one input c (or P_{Bad}) and the other one compatible with the output of NOR-9. Again, a new list, say *list-C*, is created containing four NOR gates with P_{Bad} as one

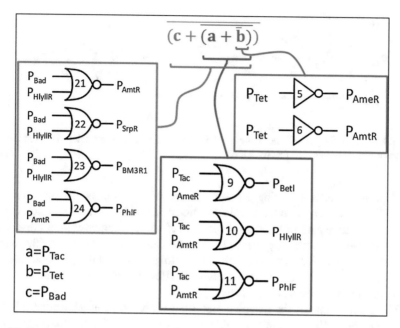

Fig. 6.9 Explanatory example of mapping algorithm

of their inputs. The second input of any of these 4 gates (in list-C) does not match with the output of NOR-9, i.e., P_{Betl}. The algorithm then steps back and removes NOR-9 from list-B, and searches for any other NOR gate compatible with the output of NOT-5, P_{Amer}. Since there are no other NOR gates available in list-B which are compatible with P_{Amer}, the algorithm further steps back and removes NOT-5 from list-A. The algorithm then selects NOT-6 and proceeds further by checking if its output, P_{Amtr}, is compatible with the available NOR gates in list-B. Both of the remaining NOR gates in list-B (2 out of 3), NOR-10 and NOR-11, match the output of the gate NOT-6. The algorithm selects NOR-10 first and proceeds ahead by checking the compatibility of its output, P_{HIyllR}, with the NOR gates available in list-C.

At this stage, there are three matching NOR gates present in the list which can be used to construct the final stage of circuit. This may result in three possible circuits constructed from the genetic gates in sequence of NOT-6 → NOR-10 → NOR-21, NOT-6 → NOR-10 → NOR-22, and NOT-6 → NOR-10 → NOR-23. There are no matching gates available in list-C with the output of second stage NOR-11 from list-B. Therefore, to implement the desired logic, the total number of circuits generated by the algorithm would be three.

While constructing genetic circuits, the algorithm avoids using the components which makes an unintended feedback loop with the preceding stage components. For example, the circuit diagram of one of the above mentioned solutions, with the components NOT-6 → NOR-10 → NOR-21, shown in Fig. 6.10, has the output of

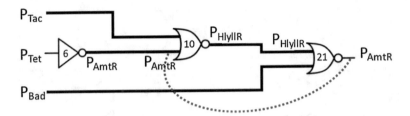

Fig. 6.10 Circuit diagram of one possible solution, for the example expression shown in Fig. 6.9, which creates an unintended feedback loop

the final gate, NOR-21, being the same as one of the input of the previous gate, NOR-10. In genetic circuits, signals are molecules and unlike electronic circuits, they do not propagate in separate wires. Therefore, it is impossible to prevent the AmtR protein generated by the NOR-21 gate from suppressing both the output promoter PAmtR for NOR-21 and the input promoter for NOR-10. Therefore, the algorithm discards this circuit for achieving the desired functionality, and hence the final number of possible circuits is two.

The pseudo code of the mapping algorithm is presented in Algorithm 6.3. The procedure, *SolveNestedEl*, in line 3 recursively extracts all the possible gates (from the genetics gates library) for each circuit element in the expression and arranges them in separate lists. Any empty list indicates that the genetic gates are not available in the library for the desired input combination. In the case of empty list, the circuit cannot be generated and the algorithm stops executing. Once the lists are generated for all circuit elements, the procedure *GatesMatching* search for the matching gates can be cascaded together. This procedure also filters out those components which forms an unintended feedback loop with the preceding gates, as described above in Fig. 6.10. Even if the list of circuit elements is not empty, it may happen that the involved gates cannot be cascaded together due to incompatible input and output. This screening is also performed by the same procedure *GatesMatching*. Once all the matching gates are found, the procedure, *GenerateCircuits*, cascades all the possible compatible gates together to construct different possible circuits. The set of generated circuits is meant to exhibit the same Boolean logic, with the combination of different genetic components, and without causing cross-interference with each other.

6.3 GeneTech Interface

GeneTech generates output circuits in the form of standard SBOL representation. For instance, one of the possible circuits, for the example shown in Fig. 6.9, consists of NOT-6 → NOR-10 → NOR-23 gates. The SBOLv diagram and the gate-level circuit schematic of this example, generated by GeneTech, are shown in Fig. 6.11a,

---

Algorithm 6.3: Pseudo code of *technology mapping*

---

input : Exp_{in}
output: Exp_{out}
// Exp_{in} = Input expression, Exp_{out} = Output Expression

1 **begin**
2   **for** *(all NOR terms, Exp_{Nor}, in Exp_{in})* **do**
3   SolveNestedEl (Exp_{Nor});
4   **if** *(any LIST is not empty)* **then**
5   **GatesMatching**; // perform gates matching
6   **GenerateCircuits**; // generate all possible circuits
7   **else**
8   *Print*"Gates not available in library;
9   **end**
1   **Procedure** SolveNestedEl (*Expression*)
2   **if** *(Exp_{Nor} contains braces)* **then**
3   **for** *(all minterms, M, in Exp_{Nor})* **do**
4   **if** *(M contains further braces)* **then**
5   *Exp* = Extract expression inside braces;
                     SolveNestedEl (*Exp*);
6   **else**
7   *Create the list, LIST, of all possible gates of M*;
8   **end**
9   **end**
10   **else**
11   **for** *(all minterms, M, in Exp_{Nor})* **do**
12   *Create the list, LIST, of all possible gates of M*;
13   **end**
14   **end**
15   **end**
16 **end**

---

b, respectively. Figure 6.11a indicates that a P_{Tet} promoter generates a protein AmtR which in turn represses the output promoter P_{AmtR}. The promoter P_{AmtR} together with the promoter P_{Tac} generates the protein HlYllR, which represses the output promoter P_{HlYllR}. The promoter P_{HlYllR} together with the promoter P_{Bad} generates a protein BM3R1, which represses the activity of the output promoter P_{BM3R1}. The promoter P_{BM3R1} is used to produce the output indicator, the yellow fluorescent protein (YFP).

The front interface of GeneTech is shown in Fig. 6.12a. The tool takes input in the form of Boolean expression specified in terms of inducers IPTG, aTc, and Arabinose.

The gates in the library [2] used in GeneTech are triggered by IPTG, aTc, and Arabinose inducers.

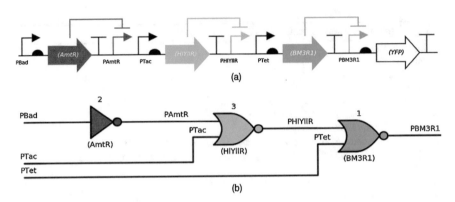

Fig. 6.11 One possible solution for the expression shown in Fig. 6.9. (**a**) Equivalent SBOLv diagram. (**b**) Equivalent gate-level circuit schematic

GeneTech generates SBOL data files for all the possible solutions for a given expression. The tool outputs those circuits only which satisfy the constraints given by a user, i.e., maximum number of gates and maximum delay offered by all the components in a circuit collectively. The generated SBOL data files can be seen in the *SBOL Data* tab as shown in Fig. 6.12b. In addition, a user can also restrict the tool to generate certain number of possible solutions. For example, a user may specify *No. of Circuits* = 2 (Fig. 6.12a) to restrict GeneTech to generate at most 2 circuits, no matter how many more GeneTech could generate for the given Boolean expression.

GeneTech also allows a user to sort the generated circuits either by the overall circuit delay or the number of gates available in each circuit, and arranges them in ascending order. Additionally, the tool also generates the output in the SBOL visual notation and the logical representation so that a user can view them both together to get an idea of the flow of a circuit in both representations. Some users, with no SBOL background, may not be able to comprehend the SBOL visual notation properly, but it becomes very easy for them to understand the logic by comparing it with the equivalent logic schematic diagram.

6.4   Experimentation by Simulation

Any Boolean expression with 3-inputs (IPTG, aTc, Arabinose) can be used for the experimentation in GeneTech.

 Once more genetic gates are developed, the gates library, in later versions of GeneTech, will be upgraded to support more variety of inputs.

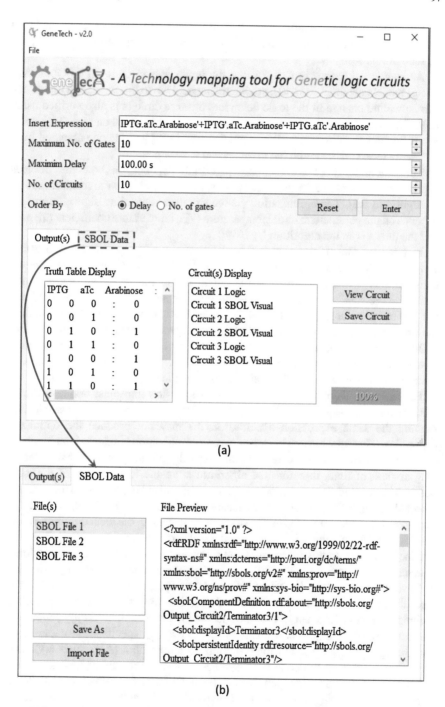

Fig. 6.12 Graphical User Interface of GeneTech (**a**) Front interface where user enters a Boolean expression along with other desired constraints. The list of circuits generated is shown under "Circuit(s) Display" window. (**b**) The SBOL data file of generated circuit(s) can be seen by clicking on the SBOL Data tab

For demonstration, experimental results of a circuit $(0 \times 0B)$ from [2] are included in this chapter. The truth table and Boolean expressions of the circuit, $0 \times 0B$, can be seen in Fig. 5.3b.

As mentioned before in Chap. 5 that the inputs B and C, in all SBOL/SBML circuit models obtained from [2], are interchanged. Due to this change in inputs, the corresponding change in the logic behaviors of these circuits is also verified using D-VASim (See Chap. 5). Therefore, the circuit $0 \times 0B$ is tested on *GeneTech* using two different expressions, i.e., the original shown in [2] and the one obtained from D-VASim (See Chap. 5) with the inputs B and C swapped.

The functionality of the mapping algorithm depends on how the Boolean expression is optimized in the first step by logic optimization algorithm (shown in Algorithm 6.1). The optimization via the simulated annealing algorithm mainly depends on the parameters: initial temperature (T_0), temperature coefficient (T_{COF}), and the time to run the algorithm (t_E).

In this experiment, the circuit $0 \times 0B$ is run for $T_0 = 10$, 20, and $30°C$; $T_{COF} = 0.90$, 0.95, and 0.99; and $t_E = 15$, 20, 25, 30, and 40 ms. For all combinations of these parameters, the circuit is run for 10 times. Based on the mean and the standard deviation values of these experiments, it was observed that the parameter values, $T_0 = 10$ and $T_{COF} = 0.90$, were suitable for generating the optimized expression. Besides these, it is understood that the optimization of a Boolean expression mainly depends on the amount of time the optimization algorithm is run for each circuit, i.e., t_E. It also depends on the combination of minterms selected randomly for optimization (see the explanation of Eq. (6.4)).

For example, the circuit may reduce the expression into most optimized form when it runs for, say 25 ms, as compared to 20 ms. It may also be possible that running the same expression again for 25 ms may not produce the optimized expression. That may happen because a wrong combination is selected randomly by an algorithm to reduce the expression. Since the algorithm is not run for sufficiently long amount of time, therefore the algorithm is unable to come out of the local optimum value and does not produce the most optimized expression. The time, t_E, reported in Fig. 6.13 is the one which ensures that the algorithm will run sufficiently long to produce the optimized expression even if it stuck in the local optimum value.

Figure 6.13 depicts the experimental results of the genetic circuit $0 \times 0B$, obtained from [2]. In this figure, separate results are included for the Boolean logic obtained originally from [2], and for the one obtained experimentally by using the logic verification method in D-VASim (See Chap. 5). The optimum time to run the algorithm, t_E, is mentioned in milliseconds, which shows that the circuit produces optimized expression when run for the specified interval of time. Exp_{Init} indicates the initial Boolean expression taken as input. Exp_{Opt} specifies the expressions after being optimized through logic optimization algorithm (Algorithm 6.1). The expressions shown in the column of Synthesis show the results of logic synthesis algorithm (Algorithm 6.2), which converts the optimized expression into NOR/NOT form. The SBOLv diagrams of all the generated circuits and their equivalent logic circuit diagram, using mapping algorithm (Algorithm 6.3), are displayed in the right most column in Fig. 6.13.

Circuit 0x0B					
Logic obtained from	Optimization			Synthesis	Technology Mapping
	t_E (ms)	Exp_{Init}	Exp_{Opt}		
Cello Paper	25	$\bar{A}\bar{B}C+\bar{A}BC+ABC$	$C(\bar{A}\bar{B}+B)$	$\overline{(\bar{C}+\overline{(B+\bar{A})})}$	1. 2. 3. 4. 5. 6.*
D-VASim Logic Analyzer	22	$\bar{A}B\bar{C}+\bar{A}BC+ABC$	$B(AC+\bar{A})$	$\overline{(\bar{B}+\overline{(\bar{A}+C)})}$	1. 2.

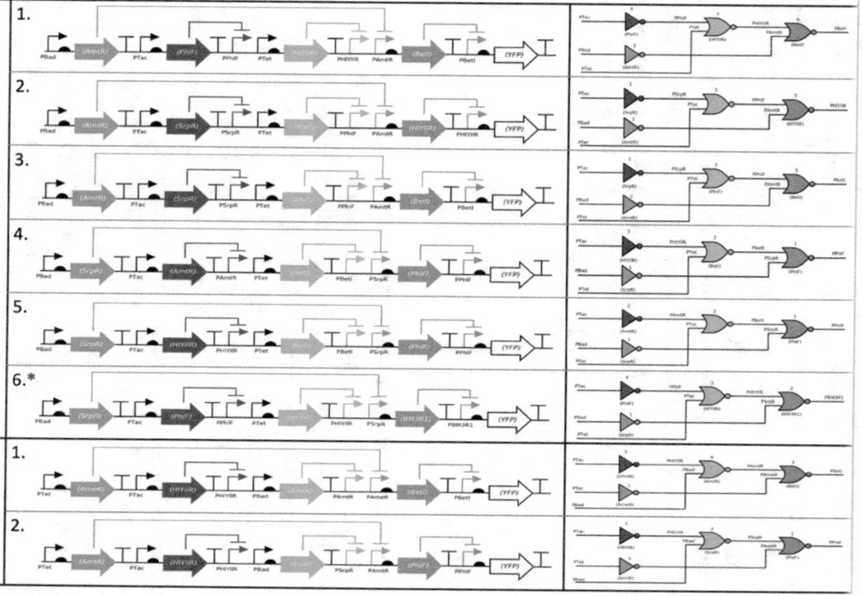

Fig. 6.13 Experimental results of *GeneTech* for the optimization, synthesis, and technology mapping of 0 × 0B circuit

For ease of understanding and to distinguish circuit components, GeneTech shows genetic gates in colors in SBOLv as well as in its corresponding logic schematic diagram. The number of genetic gates in these auto-generated circuits varies from 3 to 6, with no repetition of same genetic gates (which generates the same output protein) in any of them. In general, the number of possible solutions is more for the circuits originally obtained from [2], and less for the Boolean behavior of these circuits obtained using D-VASim (See Chap. 5). In other words, the number of possible circuits is reduced when the inputs B and C are swapped (see Problems at the end of this chapter). Among several possible solutions which GeneTech proposed for 0 × 0B circuit in Fig. 6.13, the solution similar to the one mentioned in [2] is marked with red asterisk (*).

6.5   Summary

In this chapter, it has been demonstrated with the help of an example that GeneTech can be used to develop a genetic circuit from a raw Boolean expression. In [2], a circuit is generated by a random search of compatible genetic gates using the simulated annealing algorithm. Since the circuit is created using the non-deterministic search, the solution may be different every time the process is executed. On the contrary, GeneTech finds all possible genetic circuits based on a deterministic approach. GeneTech is also scalable, meaning that it can process newly added genetic gates without any further modifications.

This tool is typically helpful for both, the biologists and the computer scientists. It is because the computer scientists do not need to learn any biological terminologies nor in which order the genetic components should be connected to construct genetic circuits. Biologists, on the other hand, do not need to learn any programming language or syntax to design genetic circuits in-silico. The users only need to specify the Boolean behavior, in the form of expression, they want to achieve in a living cell, and GeneTech lists down all the possible circuit structures to achieve it.

Problems

6.1 In Problem 5.1(ii), you evaluated the Boolean expressions of 9 different circuits obtained from [2] in the form of inputs A, B, and C. For the circuits 6–7, i.e., x70, xC4, xC8, and x0E, replace the input variables with inducers IPTG, aTc, and Arabinose, respectively.

6.2 For all the four expressions obtained in Problem 6.1, generate the genetic circuits using GeneTech for the constraints shown in Fig. 6.12.

6.3 Repeat Problem 6.2 for the Boolean expressions obtained (for circuits x70, xC4, xC8, and x0E) in Problem 5.2(ii) with inputs B and C swapped.

6.4 Compare the output circuits generated by GeneTech in Problems 6.2 and 6.3. (a) What is the general impression? (b) Why no circuit is generated for x70 when inputs B and C are swapped in Problem 6.3? (c) How come the circuits 1 and 2 for xC8, generated in Problem 6.2, are different?

6.5 What is the difference between the outputs generated by GeneTech and those present in [2] for circuits xC8 and x0E? Are both circuits equivalent? Which output is the optimized and what is the advantage of circuits produced by GeneTech?

References

1. B. Bartley, J. Beal, K. Clancy, G. Misirli, N. Roehner, E. Oberortner, M. Pocock, M. Bissell, C. Madsen, T. Nguyen, Z. Zhang, J.H. Gennari, C. Myers, A. Wipat, H. Sauro, Synthetic biology open language (SBOL) Version 2.0.0. J. Integr. Bioinform. **12**(2), 272 (2015)
2. A.A.K. Nielsen, B.S. Der, J. Shin, P. Vaidyanathan, V. Paralanov, E.A. Strychalski, D. Ross, D. Densmore, C.A. Voigt, Genetic circuit design automation. Science **352**(6281), aac7341–aac7341 (2016)
3. C.J. Myers, *Engineering Genetic Circuits* (Chapman and Hall/CRC, New York, 2009)
4. G.D. Micheli, Synthesis and optimization of digital circuits, in *Synthesis and Optimization of Digital Circuits* (1994), p. 576
5. P. Färm, *Integrated Logic Synthesis Using Simulated Annealing*, PhD thesis (2007)
6. S. Kirkpatrick, C. Gelatt Jr, M. Vecchi, Optimization by simulated annealing. Science **220**(4598), 671–680 (1983)

Part III
GDA Tool Development: A Quick Project-Oriented Approach

In this part, you will be having a hands-on experience of developing some portion of computational biology tools by first learning a graphical programming language, followed by doing projects and solving the extended challenges. This section consists of three chapters.

Chapter 7
Learn to Program Rapidly with Graphical Programming

Textual-based programming languages like C, Python, Java, etc. are all well-known and used primarily for software development. Because of their complex syntax, it is often harder for non-CS students to build up their interest in programming. It is a famous proverb that *a picture is worth thousand words*, and imagine how intuitive would it be if a user draws a program rather than writing it. Such programs are called graphical programs and the language is commonly referred to as *graphical or G* programming language. G programming language is similar to any other traditional programming language except that instead of writing lines of instructions, user *draw* a code to execute the desired logic.

In this chapter, we will study a widely known graphical programming language named LabVIEW (Laboratory Virtual Instrument Engineering Workbench). It is often a misconception that LabVIEW is only intended for test, control, and measurement applications, and that it is not a programming language like other general-purpose programming languages. LabVIEW, however, is indeed a programming language which allows a user to create a logic by wiring graphical icons together on a programming canvas. The logic diagram is then compiled to a machine code which runs on a computer processor.

G programming is equipped with the same traditional programming concepts that exist in any other programming language. For instance, it includes all the standard programming constructs including loops, conditional statements, data types, event handling, object-oriented programming, etc. The best part of LabVIEW programming platform which makes the logic development and testing intuitive is that a graphical user interface development is very easy in comparison to other traditional programming languages. LabVIEW program is called *Virtual Instrument or VI* and is saved as *.vi* extension.

Electronic Supplementary Material The online version of this chapter (https://doi.org/10.1007/978-3-030-52355-8_7) contains supplementary material, which is available to authorized users.

In this chapter, you will be introduced with the basic components of LabVIEW VIs, functions, and basic programming constructs. The main aim of this chapter is to give you a kick start to G programming, and thus not all LabVIEW functions are covered here. More information can be obtained from the resources mentioned in References. Furthermore, we assume that readers are already familiar with very basic programming concepts.

Note The trial version of latest LabVIEW can be downloaded from https://www.ni.com/en-us/shop/labview.html. The examples include in this chapter are created using MAC OS, which makes no difference if they would have been be created on Windows or Linux OS.

7.1   Getting Familiar with LabVIEW Interface

As mentioned above, LabVIEW program is referred to as a virtual instrument (VI) which comprises two different panels—a front panel and a block diagram, as shown in Fig. 7.1. Front panel acts like a graphical user interface containing *controls* and *indicators*. *Controls* are knobs, push buttons, slide switches, etc. which are used to interact with VI. That is, these are the interfaces through which inputs are given to a LabVIEW program. *Indicators* on the other hand are the components which are used to indicate the output of a G-program. Indicators include graphs, LEDs, numerical outputs, etc. The reason why LabVIEW programs are referred to as virtual instruments is because the components available to create a front panel make a LabVIEW user interface imitate physical instruments such as oscilloscope.

An example LabVIEW program for tank simulation is shown in Fig. 7.2. This figure shows that the program consists of two different windows—the upper one, shown as (a), is the front panel which contains the control knob, numeric indicators, graph, LED, etc. The panel shown as (b) at the bottom is the block diagram which contains the graphical code of tank simulation program. It contains the lower-level

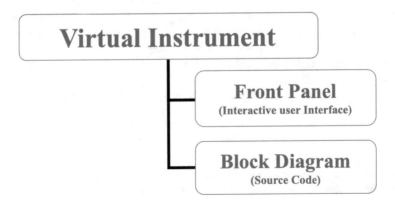

Fig. 7.1 Every virtual instrument consists of a front panel (user interface) and a block diagram (source code)

VIs (analogous to sub-routines or functions in textual-based languages), built-in functions, constants, program control structures, etc. For every control or indicator on front panel, there is a corresponding terminal on the block diagram which acts as a bridge to transfer the data back and forth between a user interface and a G-code.

Another important thing on the front panel and block diagram is the toolbar, shown at the top in Fig. 7.2a and b. Out of all the options shown, four important options (common to both front panel and block diagram) from left to right are **Run**,

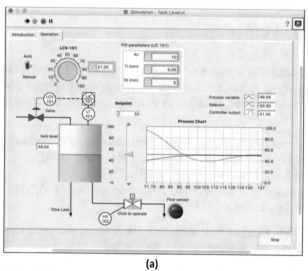

(a)

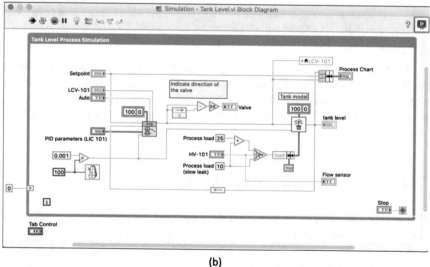

(b)

Fig. 7.2 A sample VI of a tank simulation program. (**a**) Front panel. (**b**) Block diagram (VI extracted from LabVIEW built-in examples)

Run Continuously, **Abort Execution**, and **Pause** buttons, respectively. Run button is used to run the code; run continuously button is to run the code continuously; abort execution button is used to force stop the ongoing execution; and the pause button can be used to pause the code execution.

In contrast to traditional approach of understanding the programming constructs first, we will learn the LabVIEW programming constructs by creating small LabVIEW programs and learn the functions alongside which are required to build the desired logic.

7.2   First LabVIEW Program

Suppose, we are given a task to develop a program which continuously takes a floating point input values from a user assuming that the input directly reflects the value of a temperature. Now let us develop the logic for the following set of conditions:

- Display the message "Caution: Temperature reaches 40 °C." when the input value goes beyond 40; else display "Temperature is under control."
- Turn on the LED when temperature reaches 70 °C
- Finish the program either when the temperature reaches 85 °C or when a user presses the **Stop** button

Run LabVIEW » go to **File** » click **New VI**. You can also press **Cmd+N** keys on MAC or **Ctrl+N** keys on Windows to open a new VI.

You may also consider creating a blank project and add all the relevant design VIs in it. But for now, we will only create a single separate VI.

When we create a new VI, we see two separate windows of front panel and a block diagram popped up on our computer screen. Right-click anywhere on the front panel to display the **Controls Palette** as shown in Fig. 7.3a. Here we can drag and drop desired controls and indicators on the front panel to create a front interface of our program.

7.2.1   Creating Controls and Indicators

First we have to create a control to let the user provide floating point input to our program. Go to *Numeric* category in **Controls Palette**, drag the **knob** control and drop it on the front panel. Similarly, select **String** indicator, **Round LED**, and a **Stop Button** one by one and drop on the front panel from *String & Path* and *Boolean* categories, respectively. Also, use the **Thermometer** indicator to indicate the user's input graphically. Once all the components are placed, resize and rearrange them to match the interface with the one shown in Fig. 7.4a.

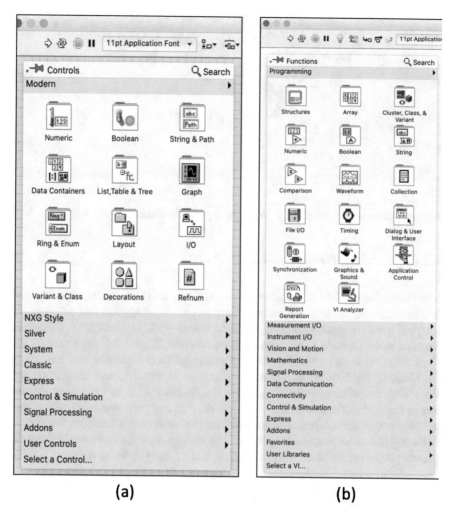

Fig. 7.3 Design palettes of a VI. (**a**) Front panel's controls palette box. (**b**) Block diagram's functions palette box

As we can see in Fig. 7.4b that four different terminals are automatically created corresponding to each control and indicator on the front panel shown in Fig. 7.4a. As mentioned before, these terminals act as data communication ports between the code and front interface. Following color conventions are used in LabVIEW for general data types:

- Pink –> Strings
- Orange –> Signed/Unsigned double/single/floating point values
- Blue –> Signed/Unsigned integer values
- Green –> Boolean

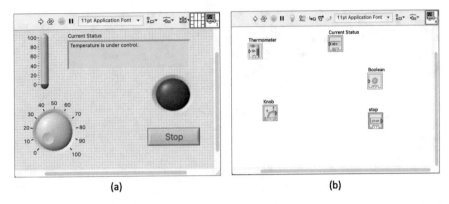

Fig. 7.4 Our designed first LabVIEW VI. (**a**) Front panel. (**b**) Block diagram

7.2.2   Building Graphical Logic

Now move on to block diagram to construct a logic to achieve the desired functionality. Remember the main three goals of the program that we need to *compare* the user input with three different values—40, 70, and 85 °C. Therefore, we need to first have a *greater-or-equal* comparator. Right-click anywhere on the block diagram to open the **Functions Palette**; go to comparison tool box; click the **Greater Or Equal?** comparator, as shown in Fig. 7.5, and drop it on the block diagram. Alternatively, we can use the *Search* option in **Functions Palette** box to find our desired function. Since we need to compare three different values independent of each other, therefore we need three such comparators. Instead of following the same procedure again, we can simply copy the comparator function already placed on the block diagram and paste it as many times as we want.

7.2.2.1   Conditional Logic: Case Structures

We also need a conditional structure to execute a code for two different scenarios i.e. display the message "Caution: Temperature reaches 40 °C." when temperature exceeds 40 °C, otherwise keep displaying the message "Temperature is under control.". In traditional programming languages, such a functionality can be implemented using *if-then-else* or *case* structures. LabVIEW provides user the **Case Structure** to implement such condition-based scenarios. The **Case Structure** has two or more subdiagrams (or cases), as shown in Fig. 7.6a. Only one conditional code executes depending on the input condition.

The case structure has two main components—**Selector Terminal** and **Selector Label**. Selector label defines the conditions to be checked depending on the data type wired at the **Selector Terminal**. If the input to selector terminal is a Boolean value, then the case structure will have only two cases—True and False, as shown in

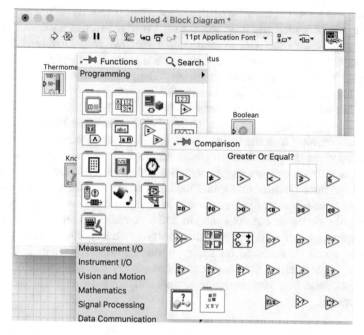

Fig. 7.5 Dropping one of the basic programming construct on block diagram

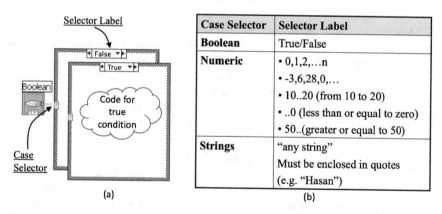

Case Selector	Selector Label
Boolean	True/False
Numeric	• 0,1,2,…n
	• -3,6,28,0,…
	• 10..20 (from 10 to 20)
	• ..0 (less than or equal to zero)
	• 50..(greater or equal to 50)
Strings	"any string"
	Must be enclosed in quotes
	(e.g. "Hasan")

(a)   (b)

Fig. 7.6 Case structure. (**a**) Boolean case structure with only two possible cases—True or False. (**b**) Chart of three different case data types

Fig. 7.6a. If a string or numeric data type is wired to the selector terminal, the case may have many sub-structures depending on the requirements. For these data types, the example conditions in **selector label** are shown in Fig. 7.6b. The important thing is to add a *Default* case that executes if the value wired at selector terminal does not match with any of the cases defined in selector label. For now, we will only stick to Boolean case structure as it fulfills our design requirements.

Search and select the **Case Structure** in functions palette and add it on the block diagram. Again, search and drop the **String Constant** from Functions Palette to block diagram inside the *case* structure under the *True* case. Double-click on the string constant to go in edit mode and write a message "Caution: Temperature reaches 40 °C.". Similarly, write a message "Temperature is under control. " in a separate string constant and place it inside **False** condition.

Next, we will start wiring the components together. Terminals can be connected by wiring their ports together. Control terminals always have their ports on the right-hand side of terminal icon and indicator terminals have it on the left-hand side. Bring the cursor toward the port of control *knob* terminal where you will notice that the cursor icon will change to a wiring spool; click and connect it to the *upper* input port of any of the three **Greater or Equal?** functions terminal. On the second terminal, we should wire a constant value, say 40, in order to compare it with the values coming from a user via control knob. To add a numeric/floating point constant to the second input port of the comparator function, we can either follow the same procedure which we followed for adding a string constant and wire it up, or we can simply right-click on the input port of comparator function and select **Create » Constant**. Similarly, wire other comparators and string constant with the *Current Status* terminal.

7.2.2.2   Continuous Iteration Using While Loops

The last design requirement says that the program should terminate either when a temperature rises above 85 °C or when the stop button is pressed. So here we need to use an **OR gate** along with a conditional structure—**while** loop. The **while** loop executes the code placed inside its boundary repeatedly until the Boolean value wired to its conditional terminal is **True**. LabVIEW checks the conditional terminal after executing the code at the end of each iteration. It stops executing if the condition is True, otherwise it runs the code again. This means that the **while** loop always executes at least once, which is analogous to the behavior of a **do-while** loop in other traditional programming languages.

Now using the functions palette again, place the OR gate on the block diagram and enclose the entire code in a **while** loop. After making appropriate connections, your final block diagram should look similar to the one shown in Fig. 7.7. The string constant statement written in the *False* case structure is stacked down and is not visible in this figure.

Figure 7.7 indicates that the thermometer is connected to the control knob, therefore the change in control knob will be directly reflected on thermometer. When the value coming through control knob (i.e. from a user) becomes equal or greater than 40, the corresponding message will be displayed. When the value goes beyond 70, the LED will turn on. When the value reaches 85 or above, the upper input condition of the **OR gate** will become *True* which in turn will produce a *True* value at the conditional terminal of the **while** loop and the program will be terminated.

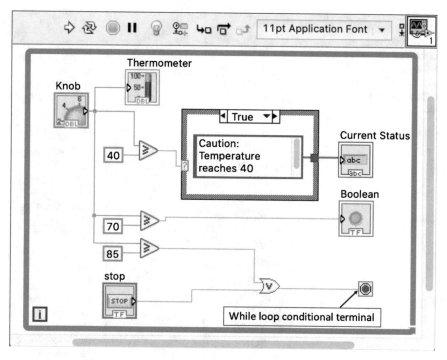

Fig. 7.7 Final block diagram (G-code) of our case example

In this example, we have covered the comparison toolbox, logic gates, conditional structures, while loop, string and double (floating point) constants.

7.3   Second LabVIEW Program

In this activity we will learn the implementation of some other important programming constructs including *Arrays, For loops, Sub VIs, Reading files, etc.*

Suppose, we have been given three different CSV (*comma separated values*) files of genetic gates. First file contains the list of NOT gates, second contains the list of NOR gates, and the third one is the main list which contains all the genetic components including promoters, RBS, genetic NOR gates, genetic NOT gates, etc. Now, our task is to make sure that the genetic circuit gates present in the first two files are all present in the third list. If any of the components from the first two lists are missing in the third one, list them down separately.

The screen captured image of the three lists are shown in Fig. 7.8, where (a) corresponds to list 1 containing NOT gates; (b) corresponds to list 2 holding NOR gates; and (c) shows the list 3 containing all genetic components.

Fig. 7.8 Three different lists of genetic components. (**a**) List of NOT gates in *List1-NOT.csv* file. (**b**) List of NOR gates in *List2-NOR.csv* file. (**c**) List of all genetic components in *List3-All.csv* file

	(a)		(b)		(c)
1	NOT1	1	NOR_1	1	PBad_P7
2	NOT2	2	NOR_2	2	PTet_P1
3	NOT3	3	NOR_3	3	NOR_R
4	NOT4	4	NOR_4	4	NOR_S
5	NOT5	5	NOR_5	5	NOT_5
6	NOT6	6	NOR_6	6	NOR_6
7	NOT7	7	NOR_7	7	NOT_23
8	NOT8	8	NOR_8	8	NOT_14
9	NOT9	9	NOR_9	9	NOR_9
10	NOT10	10	NOR_10	10	NOR_24
11	NOT11	11	NOR_AB	11	NOR_AB
12	NOT12	12	NOR_CD	12	NOR_CD
13	NOT13	13	NOR_EF	13	NOR_EF
14	NOT14	14	NOR_A	14	NOR_OL
15	NOT15	15	NOR_15	15	NOR_15
16	NOT16	16	NOR_P	16	NOR_P
17	NOT17	17	NOR_Q	17	NOR_Q
18	NOT18	18	NOR_R	18	LacI_P
19	NOT19	19	NOR_S	19	TetR_P
20	NOT20	20	NOR_T	20	NOR_T
21	NOT21	21	NOR_QW	21	NOR_QW
22	NOT22	22	NOR_22	22	NOR_22
23	NOT23	23	NOR_23	23	NOR_23
24	NOT24	24	NOR_24	24	IPTG
25	NOT25	25	NOR_25	25	NOR_25
		26	NOR_OL	26	PAmtr
		27	NOR_KJ	27	NOR_KJ
				28	PBad_yt
				29	PAmer_t
				30	PHlyIIR

(a)   (b)   (c)

The sample CSV files, shown in Fig. 7.8, are provided in the supporting material accompanied with this book.

7.3.1 Reading a File in LabVIEW

In order to compare the data of different files, we first need to read the files. The best thing about LabVIEW is its powerful built-in functions library which allows a user to develop programs rapidly as compared to other textual programming languages.

Create a New VI and name it *DataExtractor.vi*. To read a file in LabVIEW, drag and drop **Read Delimited Spreadsheet.vi** from the File I/O functions palette box. Alternatively, we can search this VI in the search bar of functions palette. Now take the cursor to the top-left port of **Read Delimited Spreadsheet.vi** » right-click » select **Create Control**.

Hover the cursor over any VI and press CMD+Shift+H (on Mac) or Ctrl+H (on windows) or Alt+H (on Linux) to open the context help about the VI or a function.

Read Delimited Spreadsheet.vi reads a specified number of lines from a file beginning at a specified character index and converts the data to a two-dimensional array of strings or double-precision or integer numbers. Since the data stored in the files, which we are reading, is of type *string*, therefore select *String* from the drop-down **Polymorphic VI Selector** option as shown in Fig. 7.9. Also, right-click on the second port at the right-hand side of the VI (*all rows*) » select **Create Indicator**.

Figure 7.9 shows how our block diagram and front panel should look like after following the steps mentioned above. The *control* we have created is shown on the front panel which will allow a user to select a file to be read. At this point, we can run the VI and specify the file (***List1-NOT.csv***) to verify if our program is able to read the file contents properly. Rename the "Read Data" array to "*List1-NOT.*"

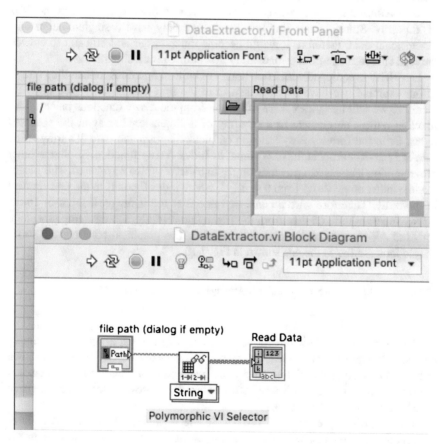

Fig. 7.9 Glimpse of a Block diagram and a Front panel after placing the *Read Delimited Spreadsheet.vi*

Turn on the *Context Help* as suggested in the previous *Tip* and hover the cursor over pink thick wire which will indicate its data type, i.e., 2D array of string (explained in the next section).

Now, to make sure that all the NOT gates listed in ***List1-NOT.csv*** file are present in the ***List3-All.csv*** file, repeat the same procedure discussed above to read the contents of ***List3-All.csv*** file, and rename the output array to "*List3-All.*"

7.3.2　Arrays and For Loops

Arrays and *For loops* are essential programming constructs of any programming language. *Array* is referred to as a set of items, of same data type, stored at contiguous memory locations. Array can be multidimensional. The built-in VI (***Read Delimited Spreadsheet.vi***) we used to read a file outputs data in the form of 2D array. Since, in our simple example, we only have a single-column one-dimensional data, therefore it would become much easier and efficient to convert the read data into one-dimensional array first before processing it.

LabVIEW is equipped with a rich library of built-in functions for handling arrays. In our example, we have to fetch data from the first column only (see Fig. 7.8), therefore we will use **Index Array** function from the *Array* functions palette. This function returns an element or a subarray of n-dimensional array at the specified index. Drop the **Index Array** function on block diagram and wire 2D array string data of *List1-Not* array at the *array* input of **Index Array** function. Similar to other programming languages, array indexing in LabVIEW is also zero-based. This means that the index range varies from 0 to $n - 1$, where index 0 points to the first element in the array. Therefore wire a constant value 0 at the *index (col)* input of **Index Array** function to fetch data from the first column of file. Do the same procedure for *List3-All* array to convert 2D data into 1D. The graphical code created so far should look similar to the one shown in Fig. 7.10.

The data wire of 1D array looks thinner as compared to that of 2D array.

Once we have the data transformed into one-dimensional array, we can simply compare the elements of two lists one by one using the *for loops*. *For loop*, as shown in Fig. 7.11 executes its subdiagram *for* the specified *n* number of times, where *n* is the value wired to the count *N* terminal. The terminal *i* provides the value of current loop iteration, ranges from 0 to $n - 1$.

To check if every element of ***List1-NOT*** file is present in the ***List3-All*** file of all genetic components, we would need to compare every single element in the first file with all the elements in the second one. This can be done in two different ways—one with two-nested loops or alternatively with a LabVIEW built-in function

Fig. 7.10 Intermediate code after converting the 2D data of both lists into 1D data

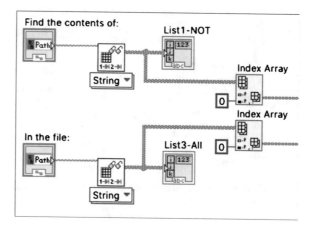

Fig. 7.11 For loop structure in LabVIEW

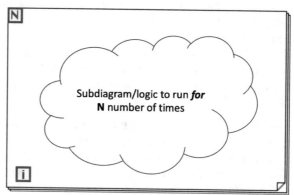

called **Search 1D Array** that allows user to find any specific element in an Array. We will utilize the power of rapid-development in LabVIEW and will use the later methodology instead of creating nested loops.

We will loop through the elements of *List1-NOT* array using *for loop* and search its indexed element in the *List3-All* array using the **Search 1D Array** function. The function returns "−1" if the specified element is not found in the array. Here we can develop a logic that if the function generates a value "−1," we can insert that element in a separate array, say *MissingNotElements* to indicate the missing elements of *List1-NOT* array in *List3-All* array.

The *MissingNotElements* array can be created and initialized as follows:

- Open *Array* palette and drop the **Array Constant** anywhere on the block diagram.
- Open *String* palette and drop the **String Constant** inside the empty *Array Constant*, which will set the type of array constant to *string*.
- Bring the cursor to the right-hand terminal of the string-array constant » Right-click » click **Create Control** to initialize the empty string array.
- Change the default name of *Array* to *MissingNotElements*.

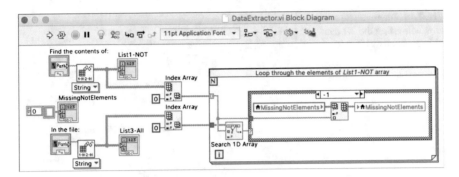

Fig. 7.12 Complete code to detect the missing elements of *List1-Not* array in the *List3-All* array

Once any element from *List1-NOT* is found missing in *List3-All*, it should be inserted into *MissingNotElements* array. The missing condition can be detected by using the *case structure*. Draw the case structure inside the *for loop*. The output of *Search 1D Array* function should be wired into the *case-selector* input of a case structure. Now, write "−1" in the *selector label* input of the case structure. Follow the steps given below to create a graphical code of inserting missing elements in the *MissingNotElements* array:

1. Search and place the **Insert Into Array** function inside the case structure.
2. Right-click *MissingNotElements* array terminal on the block diagram » go to **Create** » click **Local Variable** and place it inside the case structure.
3. Right-click on the local variable of *MissingNotElements* array » select **Change To Read** and wire it to the input *array* of **Insert Into Array** function.
4. Repeat step 2 to create another local variable of *MissingNotElements* array and wire it with the output of **Insert Into Array** function.
5. Lastly, connect the 1D wire (of *List1-Not* data) to the *new element/subarray* input of **Insert Into Array** function.

Local variable serves the same purpose as the variables do in other traditional programming languages—first initialized and then use anywhere in the program. We also initialized the string array (*MissingNotElements*) first and then call the variable later in the program to insert any new (the missing) element in it. **Insert Into Array** function inserts an element or a subarray into the existing array at the location specified by its *index* input. Since we did not specify any value to the index input, therefore every new element will be appended at the end of the list. The code created so far should look similar to the one shown in Fig. 7.12.

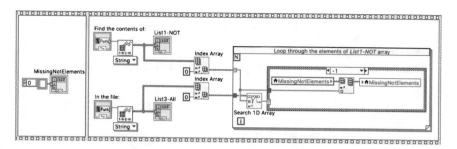

Fig. 7.13 Flat sequence structure to enforce the initialization of *MissingNotElements* array first

7.3.3   Sequence Structures

The initialization of empty array *MissingNotElements* in Fig. 7.12 indicates that the program, though, initializes the array before entering into the *for loop* but we cannot predict when will it be executed exactly—before reading the two files and/or transforming the 2D data into 1D or afterwards. Although, in this situation, it does not matter whether the array *MissingNotElements* initializes before or after reading the files, but as same as textual-based languages execute the lines of instructions in sequence, sometimes we would want to force a certain G-code to be executed in a specific sequence as well.

Let us assume that we would want to initialize the *MissingNotElements* array first and then execute the remaining code. We can do it by using the **Flat Sequence Structure** that allows user to enforce the code to be executed in a specified sequence. To do this, go to *Structures* under *Functions palette* » select **Flat Sequence Structure** » drag the cursor over the entire code to enclose it inside the **Flat Sequence Structure**. Once the code is enclosed in the structure, right-click on the left-side boundary of the sequence structure and select **Add Frame Before**. You will see another sequence structure is extended toward left. You can expand its size by dragging its boundaries toward any direction. For now, drag the left-hand side boundary to expand it horizontally enough to enclose the required logic in it. Now select *MissingNotElements* array along with the *string-array constant* wired to it and drag it to the left frame of the sequence structure, as shown in the Fig. 7.13.

Data in **Flat Sequence Structures** flow from left to right frames. In this example, first the *MissingNotElements* is initialized with empty string array, then the remaining logic is executed in the next frame.

It is very often that numerous blocks of code require to be executed in sequence, and having them enclosed in a *flat sequence structure* lengthens the code horizontally. Instead of having frames of sequences placed horizontally, we can pile them up as a *stack* so a user sees only one frame at a time. The **Stacked Sequence** structure executes frame 0 first, then frame 1, and so on until the last frame executes.

To convert the *Flat Sequence* structure shown in Fig. 7.13 into a **Stacked Sequence** structure, right-click anywhere on the boundary of **Flat Sequence**

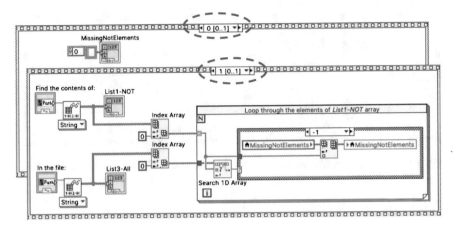

Fig. 7.14 Stacked Sequence structure containing two sequences. Though the frames are stacked up, but both the frames are shown here separately for clarity

structure » select **Replace with Stacked Sequence**. The *flat* sequence structure will be replaced with a *stack* sequence structure. Figure 7.14 shows that the stack contains two sequences indicated by **[0..1]** i.e., the stack contains sequences from 0 through 1. If there would be 10 sequences, **[0..9]** would be shown on the selector label. It can be clearly seen in Fig. 7.14 that the initialization of array, in the zeroth frame, will be executed first and then the remaining logic in the first frame. You can explore additional options of stack sequence structure by right-clicking anywhere on its boundary.

Our graphical code to check the presence of each component of *List1-Not* array in *List3-All* array is complete. We also have to make sure that the components of another list, *List2-NOR* array, are also present in *List3-All* array. Now the question is do we need to repeat the whole procedure again which we have done so far or can we just copy and paste the same code for this purpose? Instead of following any of these approaches, more appropriate way to proceed is to make the code *reusable* and use it as a *subVI*.

7.3.4 Creating a SubVI

A subVI is similar to a subroutine or a function or a method in text-based programming languages. LabVIEW allows us to create smaller sections of code, called subVIs, and call them from within another VI. We can use a subVI by dropping its icon on the block diagram of another VI, as same as we did for reading the data from file by using *Read Delimited Spreadsheet.vi* as shown first in Fig. 7.9. To make a subVI for reading and comparing two files, let us save the current working VI with different name. Go to **File » Save As »** select **Open additional copy** radio button » click **Continue...** » change the name to ***DataExtractor-SubVI*** and click **Save**.

Now we would want a user to provide file paths, of two files to be compared, as an input and get the list of only missing elements as an output. Since the code is already developed and functioning properly we do not need to make any changes except to modify the names of inputs and outputs to make the code reusable in another VI.

  Make sure that you are currently working in ***DataExtractor-SubVI.vi*** file.

Open frame 1 of stack sequence structure in the block diagram » right-click on *List1-Not* array terminal icon and select **Hide Indicator**. This action will hide the indicator from front panel. Go to front panel and verify that the array will not be visible on the front panel. Do the same for *List3-All* array to hide it from the front panel as well. Now open frame 0 and rename *MissingNotElements* to *MissingElements*. We rename it because we want to use this subVI generally to hold the missing elements, of any file, in the *List3-All* file.

Now we have to make the input and output ports of this subVI. Go to front panel and right-click on the connector pane at the top-right corner of the VI, (as shown in Fig. 7.15a) » select **Patterns** » select the fifth pattern in the first row having three blocks—two small on the left and one big on the right-hand side, as shown in Fig. 7.15b. Each rectangle on the connector pane represents an input and output terminals.

Once a pattern is selected, we must assign front panel controls (inputs) and indicators (outputs) to each of the connector pane terminal. Now click on the top-left square box of a connector pane (the box will turn black) and then click on the control named *Find the contents of:*. This will connect the control with one of the input ports of this subVI. Do the same for bottom-left box and connect it with the other file path control named **In the file:**. Finally, connect the right box of a connector pane with the output of the subVI named *MissingElements*. The connector pane and the front panel of the subVI should look like the one shown in Fig. 7.15c.

  It is recommended to organize the inputs of a subVI on the left and the outputs on the right-hand side.

7.3.4.1   Creating an Icon of SubVI

It is also a good practice to customize subVI's icon to replace it with the default icon shown at the top-right corner of the front panel. Right-click on the default icon » select **Edit Icon...** to open the *Icon Editor*. You may find built-in templates

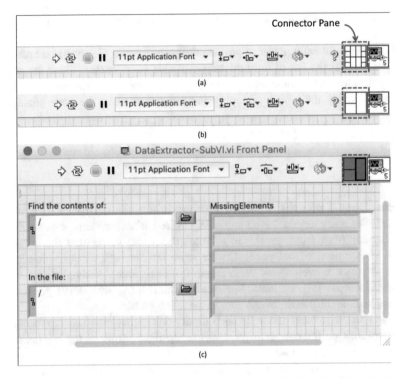

Fig. 7.15 Connector pane on the front panel of subVI. (**a**) Default view. (**b**) After selecting the pattern for two inputs and one output. (c) After connecting the pane with input and outputs

Fig. 7.16 Customized icon for the SubVI created by the authors of this book

and glyphs to construct a meaningful icon for your subVI. The icon editing tool is almost similar to the MS Paint tools on Windows OS platform. Feel free to draw any meaningful icon for your VI. In this book, we used built-in glyphs and text to give a self-expressing icon to the subVI, as shown in Fig. 7.16.

7.3.5   Putting It Altogether

The logic, required to determine the existence of the components of *List1-Not* and *List2-Nor* files in the third file *List3-All* containing all genetic components, is ready to be deployed as a subVI in our main VI. Create a new VI and save it as *CompleteDataExtractor.vi*. Press **Cmd+E** (Ctrl+E on Windows OS) to open the

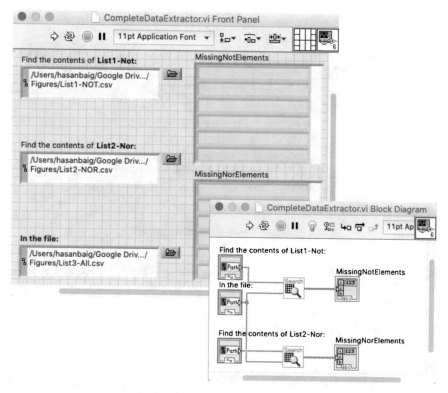

Fig. 7.17 Front panel and block diagram of the complete logic

block diagram » right-click to open the functions palette » select **Select a VI...** » choose *DataExtractor-SubVI.vi* » click **Open** » drop it on the block diagram. Alternatively, you can drag the subVI file from the file explorer and drop it on the block diagram.

Now create *controls* and *indicators* as same as we created before. Copy and paste the subVI and wire its *controls* and *indicators*. We do not need to create two separate controls for the **In the file:** terminal. Since we are checking the components of two different files in the same file so we can simply wire a single *control* to both of the subVIs. You can modify the labels of controls and indicators here in the main VI.

Front panel and block diagram of the complete logic is shown in Fig. 7.17. It can be seen that the top-level graphical code is neat and compact. To view the graphical code of SubVI, double-click on its icon to view the VI. When you run the VI, you will see the list of missing elements in both *List1-Not* and *List2-Nor* arrays in their respective output array indicators.

In this second example LabVIEW program, we have a lot many new and important concepts which are required in the software development. These concepts include reading a file, using for loops and arrays, creating sub VIs, using sequence structures, etc.

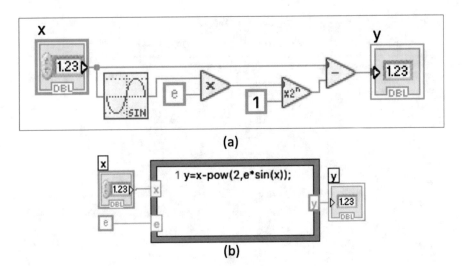

(a)

(b)

Fig. 7.18 Code to implement the mathematical expression $x - 2^{e\sin x}$. (**a**) Graphical code. (**b**) Graphical + textual code

7.4   Other Important Structures

Before we finally conclude this chapter, it is important to discuss two more important structures which are very useful in software development and will be used in the project in Chap. 8. These two structures are briefly described in the following subsections.

7.4.1   Formula Node

Sometimes it is faster to write a textual code than constructing a diagram to achieve the desired logic. For example, it might be too tedious to draw a code for evaluating the expression, $x - 2^{e\sin x}$, compared to writing a single line of code in textual-based programming languages.

The graphical code to evaluate the expression, $x - 2^{e\sin x}$, is shown in Fig. 7.18a. In contrast, a **Formula Node** structure allows us to combine the graphical code with a customized textual logic. It can evaluate mathematical formulas and supports various expressions and built-in math functions supported in C language. The equivalent *Formula Node* implementation of the graphical code, shown in Fig. 7.18a, is shown in Fig. 7.18b.

    Formula node also support basic programming structures like if-else; for, while, and do-while loops; switch-case, etc.

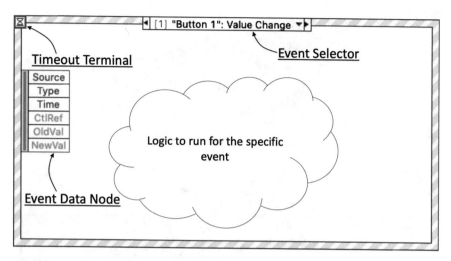

Fig. 7.19 Basic Event Structure

Figure 7.18b shows the *Formula Node* (gray squared-box) having two inputs, x and e, and an output y. The inputs/outputs in a formula node can be added by right-clicking the left/right boundaries and select **Add Input/Add Output**, respectively. The *control* terminal "x" and the constant "e" are wired to the inputs, x and e, of the formula node, respectively. Similarly, the output, y, of the formula node is connected to the terminal of y *indicator*. Figure 7.18b demonstrates how a big graphical code can be quickly represented in a single line of code.

7.4.2   Event Structure

Events are caused by user actions, for example, clicking a specific button on a graphical user interface to perform a certain task. An **Event Structure** is available to handle events in an application. Similar to a **Case structure**, you can also add multiple events to the same **Event structure** and configure them separately. When any of the event occurs, LabVIEW executes the code specified in that particular case. Figure 7.19 shows the basic event structure and its components.

Every **Event Structure** has a default (but optional) *Timeout* case to wait for the event to occur. To specify the time for an **Event Structure** to wait (in milliseconds) for an event, a numeric value can be wired at the input of **Timeout** terminal, shown in Fig. 7.19. The default timeout value is −1, which means never to time out. **Event Data Node** holds the data, returns by LabVIEW, when an event occurs. **Event Selector** allows a user to scroll through different event cases. A user may right-click anywhere on the structure's boundary to add, edit, delete, or duplicate event cases. When an event occurs, the code defined inside that specific case is executed.

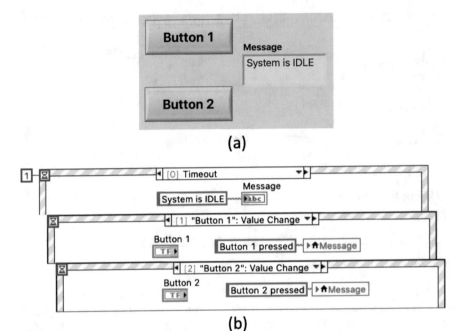

Fig. 7.20 Basic event program

Now let us create a simple program to display a different message on the same indicator when different buttons are pressed. Figure 7.20a shows the front panel having two buttons, *Button 1* and *Button 2*, and a *Message* indicator to tell the status of the program. When there is no activity on the front panel, the message "System is IDLE" would be displayed. Figure 7.20b shows three different cases of event structure. First the timeout case which displays the default message when there is no activity. Next, two different cases, separate for *Button 1* and *Button 2*, are defined. To add an event and define its property, follow the steps mentioned below:

1. Right-click on the structure's boundary and select **Add Event Case....** Another window will pop-up to edit the event as shown in Fig. 7.21.
2. Select **Button 1** from *Event Sources* box, and then select **Value Change** in the *Events* box.
3. Click **OK** to finish editing the event.

Follow the same procedure to add an event case for *Button 2*. These event cases will run when the value of a specific button is changed. Now, to display different messages on the same indicator, we have to wire the specific messages to the local variables of the *Message* terminal. Select the *Message* terminal in the *Timeout* case on block diagram » right-click » **Create** » **Local Variable** and place them inside *Button 1* and *Button 2* event cases. Wire the appropriate messages to the **Message** local variables as shown in Fig. 7.20b.

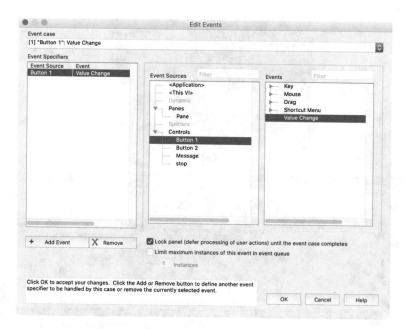

Fig. 7.21   Edit event window

As shown in Fig. 7.20b, this event structure will wait for 1ms for any event to occur and then finish. In order to run it continuously, we may consider enclosing it in the *While* loop.

 If you choose to leave the timeout terminal unchanged, set it to its default value (−1), all other logic (in parallel to event structure) will get freeze.

When you run the program, despite of pressing different buttons, you may not be able to observe any change in the *Message* indicator. This is because the program is running so fast that you cannot observe the change. We can include a *delay* function in the event cases of each button to let us observe the displayed message. Search and place **Wait (ms)** function inside both the event cases of *Button 1* and *Button 2*. This function waits for the specified value (in milliseconds) before executing the next logic. Wire the value 1000 at its input terminal in both the cases to let the message display for 1000 ms (or 1 s) before turning back to its default value defined in the *Timeout* case.

When you run the program again, you will see a message displayed on the screen. You might have noticed that the message display for 2 s with a small glitch in between. This is because the event structure is set to be executed whenever the value of a button is changed. So when we press the button, its value changes to TRUE and the corresponding message appears on the screen. When we release the button, the value changes again to FALSE and thus the event case executes again. To

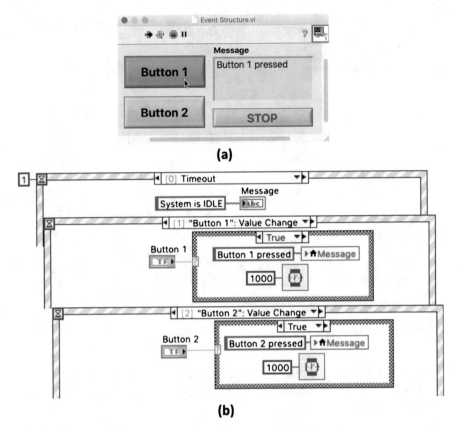

Fig. 7.22 Final updated program to capture user events on the front panel

display the message only once, we may add a **Case Structure** to display the message only when the value of a button is TRUE. The final updated program is shown in Fig. 7.22. Figure 7.22a has been captured during the run-time which indicates the moment when **Button 1** was pressed.

7.5 Summary

In this chapter, we have studied some basic concepts of graphical programming language called LabVIEW. We covered data types, loops, conditional statements, sequence structures, reading files, creating subVIs, and designing the subVI icon. We further experienced working with *controls* and *indicators* and also observed how the graphical user interfaces are developed in LabVIEW. Finally, we discussed two other important structures—*formula node* and *event structure* and learnt how to use them. In the next chapters, you will use these concepts to practice further by developing small projects.

Problems

Download free version of LabVIEW 2019 (or the latest version) from https://www.ni.com.

7.1 Create a LabVIEW program which takes two numeric inputs, x and y, and checks which one is greater and smaller. Include two LED indicators. Turn on the red LED if x is less than y, or the green one otherwise. Make your program run until the **Stop** button is pressed.

 The color of front panel objects can be changed using the *Tool Palette*. Go to **View » Tool Palette**.

7.2 Plot natural logarithmic values from 1 to 50 on the **Waveform Chart** indicator.

7.3 **Listbox** indicator is used to list the items in a user-friendly manner. All the items in different rows produce a different numeric value when selected. First row is the zeroth row which produce the value 0, row 1 produce 1, and so on. Construct a program which takes two numeric inputs, A and B, and perform the following operation listed in the listbox.

- ADD
- SUBTRACT
- MULTIPLY
- DIVIDE
- ADD+INCREMENT

The data type of **Result** indicator should be *double*, and the program should run until the **Stop** button is pressed.

7.4 Take four input Strings from the user and output one of them based on selection inputs through two Boolean Switches, S1 and S2. For example if user turns on S1 and leaves S2 off, the selection input is taken as 10 (Binary) and the system outputs the third input string.

7.5 Take any string as an input and extract the first and last character in it. Concatenate the extracted character and display as an output.

Note String size has no limitation. You may need to use several string functions. Explore the strings functions palette to achieve the desired logic.

Chapter 8
Project 1: Stochastic Simulations

In previous chapter, you learnt how to code graphically. Now, in this chapter, you will learn how to implement the Gillespie's stochastic simulation algorithm (SSA) using G programming language—LabVIEW. You will be taught how to implement a simple irreversible reaction which, later on, you can extend further to simulate more complex genetic circuit models having multiple reactions.

Stochastic simulation is used to capture the random behavior of a model with a system of variables that can change randomly (stochastically) with individual probabilities. There are several algorithms to capture the stochastic behavior of a model. Discussion of all these algorithms is beyond the scope of this book, so we will discuss only the Gillespie's Stochastic Simulation Algorithm (SSA) which is also implemented in DVASim.

8.1 Gillespie's Stochastic Simulation Algorithm

Gillespie algorithm is used to simulate reactions within living cells, where the number of reagents is low and thus the tracking of chemical interaction of different molecules is computationally feasible. Having numerous species in a system, the algorithm mainly based on the randomness of the following two aspects:

- What reaction will be executed next, i.e., which molecules of two different species combine together to form another compound or disassociate into the molecules of different species.
- When will the next reaction occur, i.e., what time will the above mentioned reaction be executed.

Electronic Supplementary Material The online version of this chapter (https://doi.org/10.1007/978-3-030-52355-8_8) contains supplementary material, which is available to authorized users.

In order to simulate a system using the Gillespie's method, the above two parameters must be selected randomly which is governed by a *propensity function* of each reaction in the system. The propensity function gives the probability, P_R, of reaction, R, occurring in time between t and $t + dt$

$$P_R dt = h_R k_R dt \tag{8.1}$$

where h_R is the number of possible combinations of reactant molecules involved in reaction, R, and k_R is rate constant.

For example, the propensity functions of the following simple irreversible reaction

$$A + B \xrightarrow{k} AB \tag{8.2}$$

are kA, kB, and kAB, where k is the reaction constant.

Gillespie developed two different methods known as the *direct method* and the *first reaction method*. Both the methods are equivalent except that the first reaction method generates a random number r_i for each reaction. In this book, we will discuss the *direct method*.

The steps to model the stochastic behavior using the algorithm is shown below:

1. **Initialization**: Initialize the current simulation time, $T = 0$, the molecular concentration of species of all reactions, and the reaction constants.
2. **Propensity functions**: Evaluate propensity functions, a_i, and also their total sum, a_{tot}, also called the *total rate of reaction*.
3. **Random Numbers**: Generate two random numbers r_1 and r_2 from the uniform distribution $(0, 1)$.
4. **Next reaction time**: The time for the next reaction to occur is given by $t + \tau$, where

$$\tau = \frac{1}{a_{tot}} \ln \frac{1}{r_1}$$

5. **Next reaction index**: The next reaction, R_j, is given by the smallest integer, j, satisfying

$$\sum_{i=1}^{j} a_i > r_2 a_{tot}$$

6. Once τ and j are obtained, the system states are updated by $X(T + \tau) = X + V_u$, where V_u is the value to be added in X; and time by $T = T + \tau$.
7. Record the updated values and jump to step 2 if the current simulation time T is less than the user-defined time to finish the simulation.

8.2   Implementation

Let us develop a program to capture the stochastic nature of the reaction (8.2) using the Gillespie's Stochastic Simulation algorithm. As discussed before, we first need to initialize the variables holding the values of molecular concentration of species, as well as the simulation time, T. At this point, we will also initialize the remaining variables shown below:

- $k1$: reaction constant
- $r1$: to hold the random value of next reaction time
- $r2$: to hold the random value of next reaction
- $a1$: to hold the propensity $k1B$
- $a2$: to hold the propensity $k1A$
- $a3$: to hold the propensity $k1AB$
- $atot$: to hold the total propensity (sum of all propensities)
- tau: to hold the value of time until the next reaction occurs

Among all the above mentioned variables, only the first one, $k1$, is of type *control* as user has to specify the value of reaction constant. We can use the **stacked sequence structure** to first declare and initialize the variables before building up the actual logic, as shown in Fig. 8.1.

Instead of initializing single variables for holding the molecular concentration of species A, B, and AB, we initialized the array for each of them to track their concentration over the time, as shown in Fig. 8.1. All the variables are declared as

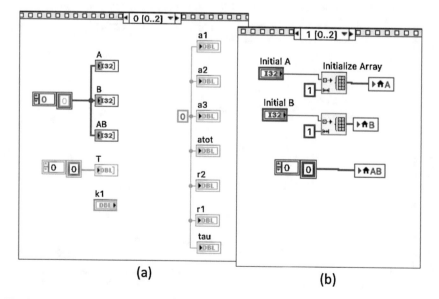

(a)   (b)

Fig. 8.1 Declaration and initialization of variables

double data type and initialized to a value 0, except the molecular concentrations which are declared as integer data types and initialized as an empty array in the 0th sequence structure (shown in Fig. 8.1a). In the next sequence structure, the molecular concentration of input species, A and B, is initialized by a user with the help of *control* terminals—*Initial A* and *Initial B*, as shown in Fig. 8.1b. It can be noticed that a new built-in function, **Initialize Array**, is used from the *array palette* to initialize these arrays.

 In the authors' provided solution, the initial default values for the concentration of inputs A and B are set to 100 and 98, respectively.

Once the initialization is done, we can start developing the logic step-wise, from the next sequence structure. Now, in the next (*2nd*) sequence structure, we would need to nest another **stacked sequence structure** to implement the algorithm step-wise. First thing we would do is to calculate the propensities and the total rate of reaction, as shown in Fig. 8.2.

Figure 8.2 indicates that the outer **stacked sequence structure** (frame 2 [0..2]) contains a nested stacked sequence which contains the code to estimate individual as well as total propensities. The nested stacked sequence structure itself is enclosed in a while loop because the process has to be repeated again until the loop's terminating condition (discussed later) is reached. It can also be noticed in the figure that the loop count terminal, i, is tracked using the indicator, L_cnt, and is used to extract

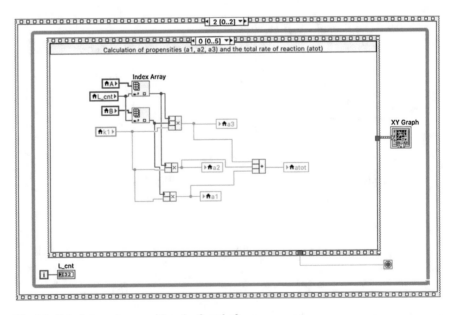

Fig. 8.2 Calculation of propensities $a1$, $a2$, and $a3$

elements from the arrays of input molecular species. For instance, at the beginning of simulation, the initial molecular concentration of species in arrays A and B will be 100 and 98, respectively. Also the value of L_cnt would be 0, therefore the current molecular concentration at 0th index of array will be extracted. Similarly, the updated concentration from the array is extracted in the next higher loop iterations with the help of L_cnt variable.

Now, we need to estimate the next reaction time, $tau(\tau)$, and the next reaction value according to step 4 mentioned in Sect. 8.1. This is shown in the next frame (1 [0..5]) of nested sequence structure in Fig. 8.3. It can be noticed that some new built-in LabVIEW functions are used including **Random Number**, **Reciprocal**, and **Natural Logarithm**.

Before we move on to randomly estimate the reaction to be executed, we first add the value of $tau(\tau)$ in current simulation time value, accessed by L_cnt in array T, and update the array by inserting the new one at the next index $L\_cnt + 1$. This is implemented in the next frame of sequence structure (2 [0..5]) in Fig. 8.4.

In the next frame (3 [0..5]) of nested stacked sequence, we implement the logic for executing the next reaction based on the value of $r2$ according to point 5 discussed in Sect. 8.1. To make the code compact, we have employed a hybrid approach of implementing the logic by integrating the textual code (using **Formula Node**) with the graphical logic, as shown in Fig. 8.5.

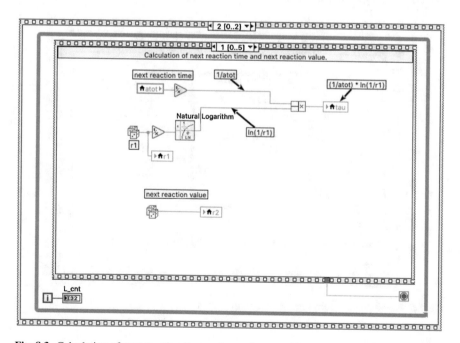

Fig. 8.3 Calculation of next reaction time and a random number generator for the next reaction index

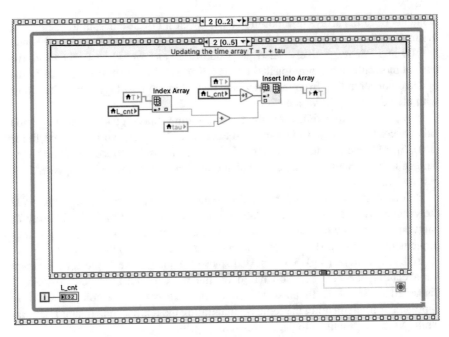

Fig. 8.4 Updating the simulation time

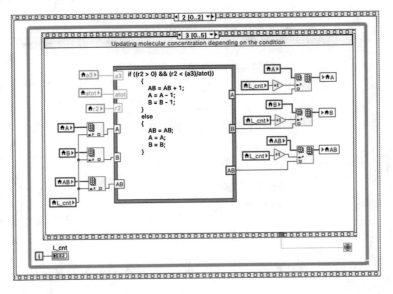

Fig. 8.5 Hybrid logic to select the next reaction to occur

Since we have only one reaction in this example, therefore we compare the propensity of the reaction's product (AB) with a product of random number, $r2$, and the total rate of reaction, *atot*. In this example, the system basically decides whether the species A and B interacts (reduced concentration count by 1) to form the product AB (increase concentration count by 1), or simply no reaction occurs (concentration remains same). This is shown in the formula node in *if* and *else* blocks, respectively, in Fig. 8.5. The textual code decides which reaction to execute, whereas the graphical code is used to fetch the current molecular concentration as well as inserting the updated values in the concentration arrays.

Now that the algorithm is implemented, we finally need to project the data graphically for a user to view and set the time to finish the simulation. This is shown in Fig. 8.6a, b, respectively. Figure 8.6a shows that the molecular concentrations of all species are first **bundled** with the simulation time, T, and then integrated in a single array using the **Build Array** function. The array of complete data is then connected to the **XY Graph** indicator to display the entire data graphically. In Fig. 8.6b, we compare current simulation time with the user-defined value,

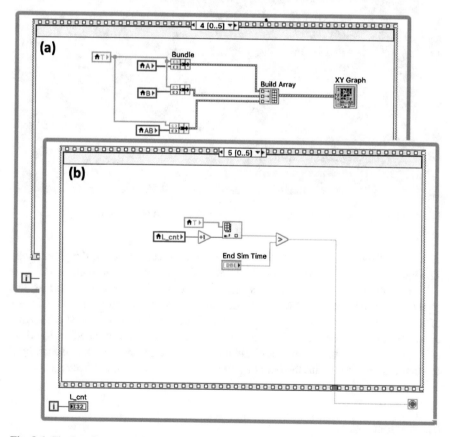

Fig. 8.6 Final settings (**a**) code to represent the molecular concentration graphically (**b**) logic to control the termination of while loop

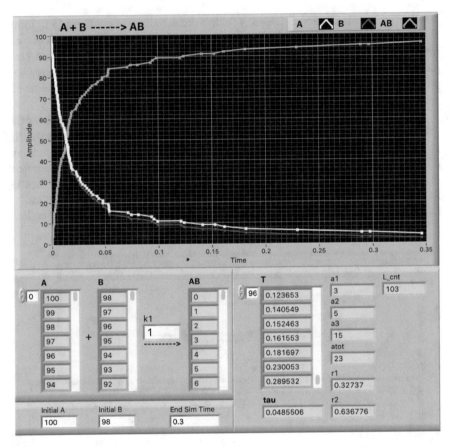

Fig. 8.7 Front panel of the stochastic simulation algorithm for the given example

End Sim Time, to decide whether to repeat the process or finish the simulation. The loop is terminated if the current simulation time exceeds the *user-defined* value.

The front panel of the code is shown in Fig. 8.7. It produces almost the equivalent trajectories every time but different values because of the random nature of algorithm. The controls *k1*, *Initial A*, *Initial B*, and *End Sim Time* allow a user to specify the values of reaction constant, initial concentrations of species A and B, and the amount of time to run the simulation, respectively. A user can also analyze other simulation data with the help of other indicators shown at the front panel.

8.3   Discussion

The example we discussed in this chapter is for a single reaction only, i.e., we could model it easily with the help of single *if-else* structure. This approach can be extended even for few more number of reactions in a system. However, to generalize the implementation for any number of reactions, the matrix-based approach of species stoichiometry is followed. In this approach, the molecular concentrations of species, to be updated in each reaction, are defined in separate rows.

Consider the same reaction (8.2) again as a *reversible* reaction as shown below:

$$A + B \underset{kr}{\overset{kf}{\rightleftharpoons}} AB \tag{8.3}$$

The reaction (8.3) can also be considered as the following two different reactions, with kf and kr being the forward and reverse reaction constants, respectively.

$$A + B \overset{kf}{\longrightarrow} AB \tag{8.4}$$

$$AB \overset{kr}{\longrightarrow} A + B \tag{8.5}$$

In the example we discussed in previous section, we assumed that the initial molecular concentration of species A and B are 100 and 98 molecules, respectively. Assuming the same molecular concentration, when the reaction, described as Eq. (8.4), is qualified to be executed using Gillespie's approach, we can say that one molecule of A and B species are combined to form one molecule of AB dimer. In other words, one molecule from the concentration of species A and B is reduced and one molecule is added in the molecular concentration of AB dimer. Similarly, with every execution of reaction (8.5), where AB dimer is decomposed into A and B, one molecule of AB is reduced and one molecule of each A and B is increased.

The stoichiometric updates of species involved in reactions (8.4) and (8.5) can be defined in matrix form as shown in Table 8.1. When any reaction (8.4) or (8.5) is randomly selected, based on point 5 described in Sect. 8.1, the corresponding vector is selected (labelled as R1 or R2 in Table 8.1) and numbers assigned to each species are added to their current concentration values. For example, when a reaction, $R2$, is selected to be executed, the molecular concentrations of species A and B are increased by 1 and that of dimer AB is reduced by 1.

To generalize the simulation for all kind of models, having different number of reactions, the program should be written to extract all species and reactions automatically from the SBML file, and generate a corresponding matrix itself.

Table 8.1 Species stoichiometry matrix of reactions (8.4) and (8.5)

	A	B	AB
R1	−1	−1	1
R2	1	1	−1

The general implementation can be observed in the source code of DVASim. Open *SSA_Template.vi* » frame 9 of stacked sequence structure » *while loop* placed on the left-hand side » *Timeout* frame of **Event Structure**.

8.4   Challenge

We discussed the functionality of Lac Operon in Sect. 2.2. The logic of Lac Operon is briefly summarized in Fig. 2.4 and the corresponding SBOL diagram is shown in Fig. 2.5. We observed that the Lac Operon transcribes three different genes—lacZ, lacY, and lacA, when glucose is absent (i.e., the concentration is below its threshold level) and Lactose is present (i.e., the concentration is above threshold level) in the cell.

To simplify the output, we can follow Fig. 2.4b which says that an output, transcription (T), appears when glucose (G) is absent and Lactose (L) is present in the cell. Now model the stochastic behavior of the Lac Operon in a similar manner we followed in Sect. 8.2. Some important aspects and the requirements are given below:

1. The front panel of the expected simulator is shown in Fig. 8.8 below which has:

   - the control knobs for both inputs—Glucose and Lactose
   - both analog and digital graphs
   - controls to specify the threshold values of each input separately (e.g., threshold value of glucose = 65 and lactose = 45 as shown in Fig. 8.8)
   - a knob to control the speed of simulation
   - a Stop button to finish the simulation at any instant of time
   - a block (highlighted with red-dotted line) which indicates the Boolean logic based on the current simulation data (Bonus)

2. The initial concentrations of glucose, lactose, and transcription should be initialized with a *random* value between 0 and 100 molecules.
3. The decrements in the molecular concentration of glucose and lactose species should also reflect on their respective control knobs.
4. User should be able to interact with the model and change the molecular concentration of inputs at any instant of time.

 Event Structure will be required to implement the execution of a code related to specific control.

5. The digital graph should plot the values 0 or 1 based on the comparison of current concentration level of each species with their corresponding threshold values.
6. Boolean expression analyzer (Bonus), which:

   - should be able to analyze the simulation data and construct the Boolean expression out of it

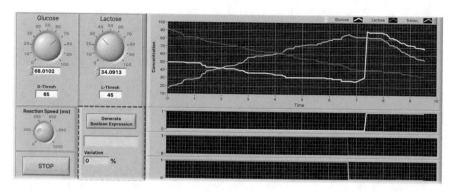

Fig. 8.8 Front interface of Lac operon stochastic simulator

- should also specify the percentage of *variation* obtained in the Boolean expression, i.e., the amount of simulation data which do not fit in the estimated expression

 Estimating the percentage variation of Boolean expression is exactly opposite to the percentage fitness discussed in Algorithm 5.4.

8.5 Summary

In this chapter, we learnt how to implement the Gillespie's stochastic simulation algorithm in LabVIEW. We further learnt how to make the simulation interactive. The solution of code challenge is available with instructors' resources.

Chapter 9
Project 2: Parsing the SBML File

In this project, you will learn how to read and parse the SBML file using G programming language—LabVIEW. We will learn how to parse an SBML file by developing a parser for few SBML components. The experience can be extended to develop a complete SBML parser.

The Systems Biology Markup Language (SBML) is an XML-based representation format for biological models including metabolic pathways, cell signaling pathways, gene expression, etc. It has been adopted as a standard for representing computational biological models in systems biology community. More details and the latest version of SBML can be accessed at http://sbml.org/Documents/Specifications.

Consider a simple network of reactions in Fig. 9.1. This set of biochemical reactions can be represented in the SBML format. The symbols with square brackets, e.g., $[S_1]$, indicate molecular concentration of species; the arrows represent the reactions; and the math above the arrows indicates the rates at which the reactions take place. Any biological model including the one shown in Fig. 9.1 can be broken down into different number of components. For example, for the model of biochemical reactions, shown in Fig. 9.1, there will be *reactant species*, *product species*, kinetic *reactions*, *parameters*, etc.

SBML allows to describe the type of component in a model by a specific type of data object containing the relevant information. Any biological model represented in SBML format may contain the SBML components, shown in Fig. 9.2, with all being optional.

Any SBML model may contain all or few of the components shown in Fig. 9.2. The detailed description of each of these components can be accessed from the SBML specification document. In the next section we will implement the graphical code to extract the SBML components from a sample SBML model.

Electronic Supplementary Material The online version of this chapter (https://doi.org/10.1007/978-3-030-52355-8_9) contains supplementary material, which is available to authorized users.

Fig. 9.1 Simple network of biochemical reactions to be represented in SBML (example taken from SBML specification document)

$$S_1 \xrightarrow{k_1[S_1]/([S_1]+k_2)} S_2$$

$$S_2 \xrightarrow{k_3[S_2]} S_3 + S_4$$

Fig. 9.2 SBML components (obtained from SBML specification document)

beginning of model definition
  list of function definitions (optional)
  list of unit definitions (optional)
  list of compartments (optional)
  list of species (optional)
  list of parameters (optional)
  list of initial assignments (optional)
  list of rules (optional)
  list of constraints (optional)
  list of reactions (optional)
  list of events (optional)
end of model definition

```
<listOfSpecies>
    <species metaid="s2" id="s1" name="LacI" compartment="default" initialAmount="10" charge="0">
        <annotation/>
    </species>
    <species metaid="s3" id="s2" name="mRNA1" compartment="default" initialAmount="0" charge="0">
        <annotation/>
    </species>
    <species metaid="s11" id="s3" name="Cat 1 (L)" compartment="default" initialAmount="1.3" charge="0">
        <annotation/>
    </species>
    <species metaid="s10" id="s4" name="Repressor Protein" compartment="default" initialAmount="0" charge="0">
        <annotation/>
    </species>
    <species metaid="s1" id="s5" name="Lac Operon" compartment="default" initialAmount="10" hasOnlySubstanceUnits="true" charge="0">
        <annotation/>
    </species>
    <species metaid="s6" id="s6" name="Cat 2 (G)" compartment="default" initialAmount="0">
        <annotation/>
    </species>
    <species metaid="s5" id="s7" name="mRNA2" compartment="default" initialAmount="0">
        <annotation/>
    </species>
    <species metaid="s4" id="s8" name="Lac Protein" compartment="default" initialAmount="0">
        <annotation/>
    </species>
</listOfSpecies>
<listOfParameters>
    <parameter metaid="k1" id="k1" name="Reaction1" value="1" units="substance"/>
    <parameter metaid="k2" id="k2" name="Reaction2" value="1" units="substance"/>
    <parameter metaid="k4" id="k4" name="Reaction4" value="1" units="substance"/>
    <parameter metaid="R1" id="k_rp" name="Repressor rate" value="1" units="substance"/>
    <parameter metaid="A1" id="kG" name="Repressor rate of Glucose" value="1" units="substance"/>
    <parameter metaid="nL" id="nL" name="nL" value="1" units="substance"/>
    <parameter metaid="nRP" id="nRP" name="nRP" value="1" units="substance"/>
    <parameter metaid="nG" id="nG" name="nG" value="1" units="substance"/>
</listOfParameters>
```

Fig. 9.3 Sample SBML file showing two different components of SBML model—list of species and list of parameters

The sample portion of SBML file is shown in Fig. 9.3 to indicate how different components of SBML model are distinguished. Figure 9.3 shows two components of a biological model—the *list of species* and the *list of parameters*. It can be noticed that these two components, list of species and list of parameters, are enclosed in separate unique tags—$< listOfSpecies > ... < /listOfSpecies >$ and $< listOfParameters > ... < /listOfParameters >$, respectively. As the name suggested, these components are used to separately hold all the species and parameters involved in the creation of a computational model.

9.1   SBML Parsing

Now having a brief information about the structure of SBML representation, we will develop a graphical code and follow a modular approach to extract model's information from an xml file. In this chapter, we will use the SBML model of a genetic AND gate (used in Part II of this book.). We will also be using some new string functions which we have not used before. Brief explanation of these functions is added with the "Idea" symbol.

First, we need to read the SBML file which can simply be done by using the **Read from Text File** function, as shown in Fig. 9.4 below. This figure shows that the **stacked sequence structure** is being used, and the necessary *indicators* are initialized in 0th frame of a stacked sequence structure (not shown). The entire contents of SBML file are read by specifying the file path in the *SBML File control* at the front panel (not shown in the figure).

   The function "**Read from Text File**" reads a specified number of characters or lines from a file. A complete file is read if the read *count* is not specified.

Once the SBML file is read in LabVIEW, we can create separate functions (Sub VIs) to extract the information about species, parameters, reactions, and mathematical expression of reaction kinetics as shown in Fig. 9.5a, b, c, and d, respectively. The *Read SBML File* indicator holds the content of SBML file, as shown in Fig. 9.4, which is then passed to separate functions (frames 2, 3, and 4) to extract the information of species, parameters, and reactions as shown in Fig. 9.5a, b, and c, respectively. The implementations of these Sub VIs are discussed separately in the following sub sections.

Fig. 9.4 Reading SBML file using *Read from Text File* function

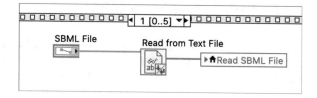

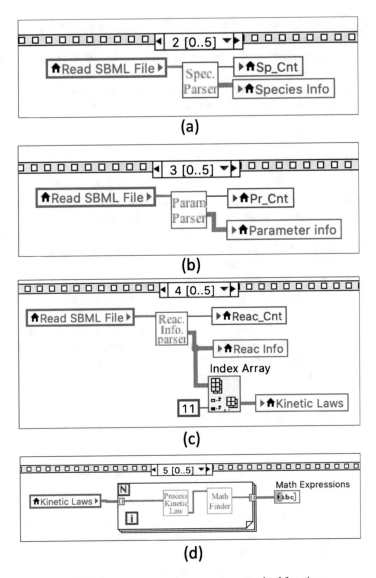

Fig. 9.5 Extraction of SBML components using separate customized functions

9.1.1   Extraction of Species Data

The sub VI shown in Fig. 9.5a takes the entire SBML file as a string input and gives the total number of species (Sp_Cnt) as well as the entire species-related information in the string array ($Species\ Info$).

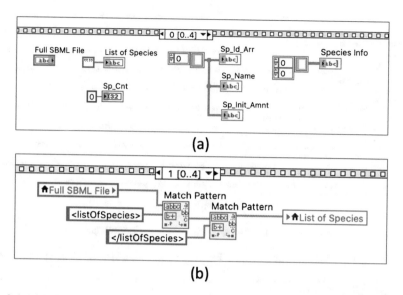

Fig. 9.6 Initial frames to extract species information. (**a**) Initialization of indicators. (**b**) Extraction of list of species

We will now develop the internal logic of the subVI shown in Fig. 9.5a. After initializing all the necessary indicators to hold the specific information (Fig. 9.6a), we first extract the data enclosed between *<listOfSpecies>* and *</listOfSpecies>* tags (Fig. 9.6b) using the **Match Pattern** string function. Two **Match Pattern** functions are used, first to extract the data after the tag *<listOfSpecies>* and then used again to extract the text before the tag *</listOfSpecies>*. In this way, the text between the two tags is extracted.

   Match Pattern function searches the specified expression (2nd input) in the input string (1st input) starting at the index specified (input terminal not shown). If the function finds a match, the input string is split into *before matched sub-string* and *after matched sub-string*.

The string extracted after this operation would look like the one shown in Fig. 9.7 below. The data stored in this portion of SBML contains several different pieces of information out of which we will extract the ID, Name, and the Initial Amount for each species used in the model.

Next we need to determine the total number of species present in the SBML model. Figure 9.7 suggests that every new species starts with a tag *<species*, we can therefore count the number of *species* tag to determine the total number of species present in the model. The program should search for the *<species* tag from the beginning of input string and then search for the next one from the offset where the previous one was found and so on until all the species are counted. There could

```
    <species  boundaryCondition="false"  constant="false"  metaid="iBioSim31"  hasOnlySubstanceUnits="true"  initialAmount="0"
compartment="Cell" id="CI"/>
    <species  boundaryCondition="false"  constant="false"  metaid="iBioSim32"  hasOnlySubstanceUnits="true"  initialAmount="0"
compartment="Cell" id="GFP"/>
    <species  boundaryCondition="true"  constant="false"  metaid="iBioSim33"  hasOnlySubstanceUnits="true"  initialAmount="0"
compartment="Cell" id="LacI"/>
    <species  boundaryCondition="true"  constant="false"  metaid="iBioSim34"  hasOnlySubstanceUnits="true"  initialAmount="0"
compartment="Cell" id="TetR"/>
    <species  boundaryCondition="false"  constant="false"  metaid="iBioSim35"  hasOnlySubstanceUnits="true"  initialAmount="2"
sboTerm="SBO:0000590" compartment="Cell" id="Promoter_GFP"/>
    <species  boundaryCondition="false"  constant="false"  metaid="iBioSim36"  hasOnlySubstanceUnits="true"  initialAmount="4"
sboTerm="SBO:0000590" compartment="Cell" id="P1"/>
    <species  boundaryCondition="false"  constant="false"  metaid="iBioSim37"  hasOnlySubstanceUnits="true"  initialAmount="2"
sboTerm="SBO:0000590" compartment="Cell" id="P3"/>
    <species  boundaryCondition="false"  constant="false"  metaid="iBioSim38"  hasOnlySubstanceUnits="true"  initialAmount="0"
compartment="Cell" id="P3_mRNA"/>
    <species  boundaryCondition="false"  constant="false"  metaid="iBioSim39"  hasOnlySubstanceUnits="true"  initialAmount="2"
sboTerm="SBO:0000590" compartment="Cell" id="P2"/>
```

Fig. 9.7 Data extracted between the tags *<listOfSpecies>* and *</listOfSpecies>*

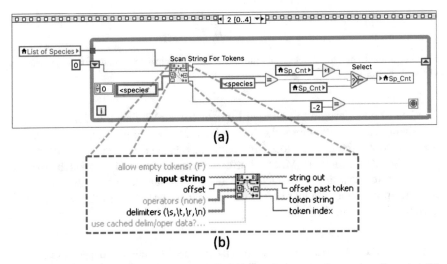

Fig. 9.8 Counting the number of species. (**a**) Graphical code to count the number of species. (**b**) Input/output diagram of **Scan String for Tokens** function

be several different methods to achieve this, but in this book we have used the built-in LabVIEW function, named **Scan String for Tokens**, inside a *while* loop in combination with *Shift registers*.

Figure 9.8 shows the graphical code to detect the total number of species in the model. Since the portion of SBML which only contains the *List of Species* has already been extracted (see Fig. 9.7) in previous frame of stacked sequence structure (Fig. 9.6), thus we can only focus on the extracted data present in the *List of Species* indicator. The local variable of this indicator is provided as an input to the function **Scan for Strings Token**, which scans for the token (*<species*), define at the input *operators* array, beginning from the *offset* value 0.

Shift registers are used to pass values from previous loop iterations to the next iterations. It appears as a pair of terminals directly opposite to each other on the vertical sides of the loop. The up-arrow terminal on the right side stores data on the completion of iteration which is then used by the down-arrow terminal on the left side in next iteration.

 The starting value of **Shift registers** should be initialized, by wiring the down-arrow terminal on the left side of loop, with some value. If not, the program will use the previous values of loop iterations, even the program reruns again, until it is completely closed and open up again.

 Select function acts as two-to-one multiplexer. It passes the value, wired to its upper terminal, forward when the Boolean input is true, otherwise it passes on the value wired to its bottom terminal.

The shift register is initialized with the offset value 0 to start searching from the beginning of *List of Species* string. **Scan for String Tokens** function searches for the token strings, defined in the *operators* array, and outputs the *offset past token*, *token string*, and the *token index*. Once the token string (*<species* tag) is found, the variable to count the species (*Sp_Cnt*) is incremented and updated with the help of *Select* function.

 Multiple tokens can be defined in *operators* array. See LabVIEW Help for more details.

When no token string (*<species* tag) is found, the function **Scan String for Tokens** returns −2 at the *token index* output which in turn terminates the *while* loop, as shown in Fig. 9.8.

Now that the number of species is determined in the model, we can run a *for* loop to extract the information we initially aimed for, i.e., *ID*, *Name*, and *Initial Amount* of species. We could have done this in the same *while* loop when we were counting the number of species, however, it is better to keep the logic simple and separate. Figure 9.9 shows how the **Match Pattern** function is used again in a *for* loop along with *shift registers* to extract the IDs, Names, and Initial Amounts of all the species in the model. The information to be extracted, for each attribute, is enclosed in the quotes as can be seen in Fig. 9.7. Therefore, two back-to-back **Match Pattern** functions are used to extract the required information and store in the respective arrays (Fig. 9.9).

We have gathered the required information related to species and stored it in three different one-dimensional arrays. For all the data to remain intact, it is better to have the information stored in a single multi-dimensional array, which is done in the last frame of *stacked sequence structure* and shown in Fig. 9.10.

 The species data obtained are not sorted in order. It would be a good exercise to arrange the species data, stored in the two-dimensional array—*Species Info*, in ascending order according to their IDs.

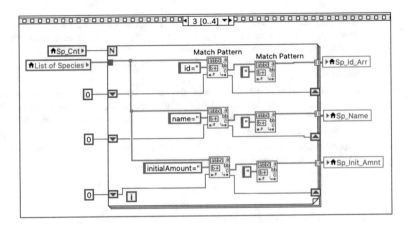

Fig. 9.9 Extracting species IDs, Names, and Initial Amounts

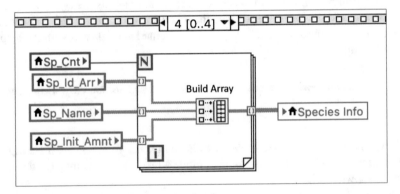

Fig. 9.10 Storing all species data in multi-dimensional array

Now we have implemented the logic to extract the species information from the SBML file, we need to use this VI as a subVI in our main program, the block diagram of which is shown in Fig. 9.5a. Now assign the inputs (*Full SBML File*) and the outputs (*Sp_Cnt* and *Species Info*) in the *connector pane* (see Fig. 7.15 in Sect. 7.3.4). Also, refer to Sect. 7.3.4.1 to create an icon for this sub VI such that it matches to the one shown in Fig. 9.5a.

9.1.2   Extraction of Parameters Data

The process of extracting the model's parameters from the SBML file is exactly same as extracting the species information. The only difference is the different tags and attributes. The data related to parameters can be acquired by using the *<listOfParameters>* and *</listOfParameters>* tags, in place of *<listOfSpecies>* tags as shown in Fig. 9.6b. Similarly, you would be required to use "<*parameter*" tag (as shown in Fig. 9.8a) to count the number of parameters used in the model.

In addition, extract the values of *id*, *name*, and *value* of parameters as same as we did for species in Fig. 9.9. Lastly, follow the same steps to store your logic to use as a separate Sub VI, as shown in Fig. 9.5b.

You may, of course, consider developing a single generic Sub VI to parse species and parameters info. In that case, you would need to make the additional input ports to specify the tags for species and parameters distinctly.

9.1.3 Extraction of Reactions Data

In comparison to species and parameters, there are many attributes associated with reactions to be extracted. For each reaction, we have to extract nearly about 12–13 attributes to make the reactions, involved in a model, completely readable to a user. The initial steps of extracting the reactions' information are similar to that of species and parameters, i.e., the extraction of reactions data between <*listOfReactions*> and </*listOfReactions*> tags, and counting the number of reactions using <*reaction* tag. This is similar to the code developed in the frames 1 and 2 of *stacked sequence structure* for species, as shown in Figs. 9.6b and 9.8a.

The data between <*listOfReactions*> and </*listOfReactions*> tags is stored in the *List of Reactions* variable. The glimpse of the portion of this data is shown in Fig. 9.11, which contains the data for a single reaction only. Because of the space limitations, the data shown in this figure is further cropped. The image is included just to have the readers an idea of the length of reactions information. It consists of several different attributes which tells us if the reaction is reversible; if it is fast; what are the reactants, products and their stoichiometries; the local parameters, presence of reaction modifiers, kinetic laws, etc.

To extract the basic information of a reaction (first line in Fig. 9.11), i.e., the ID, if the reaction is fast, and if it is reversible, we can follow the similar code which we used for species in Fig. 9.9. However, there would be a slight difference in extracting the reactions' IDs. There are multiple *id*=" " fields present inside every reaction, as shown in Fig. 9.11. These *id* fields represent the ID of reaction and the IDs for local parameters. Therefore, in order to extract the reaction's ID, we cannot simply look for the *id*=" " field in every iteration of a loop. We have to guide the search methodology to look for the reaction ID only and then jump to the next "<*reaction*" ID instead of scanning further information of the same reaction.

This can easily be implemented by looking at the <*reaction* tag first in every iteration of a loop, as shown in Fig. 9.12. When the tag is found, program looks for the *id*=" " field which would be the first reaction ID. In the next iteration, the

```
70    • • • •
71    <reaction compartment="Cell" fast="false" id="R_abstracted_production_P1" reversible="false">
72      <listOfProducts>
73        <speciesReference stoichiometry="10" constant="true" species="CI"/>
74      </listOfProducts>
75      <listOfModifiers>
76        <modifierSpeciesReference species="LacI"/>
77      </listOfModifiers>
78      <kineticLaw>
79        <math xmlns="http://www.w3.org/1998/Math/MathML">
80          <apply>
81            <divide/>
82            <apply>
83              <times/>
84              <apply>
85                <times/>
86                <ci> ko__P1 </ci>
87                <ci> ng__P1 </ci>
88              </apply>
89              <apply>
90                <times/>
91                <ci> Ko__P1 </ci>
92                <ci> RNAP </ci>
93              </apply>
94            </apply>
95            <apply>
96              <plus/>
97              <apply>
98                <plus/>
99                <cn type="integer"> 1 </cn>
100               <apply>
101                 <times/>
102                 <ci> Ko__P1 </ci>
103                 <ci> RNAP </ci>
104               </apply>
105             </apply>
106             <apply>
107               <power/>
108               <apply>
109                 <times/>
110                 <ci> Kr__LacI_P1 </ci>
111                 <ci> LacI </ci>
112               </apply>
113               <ci> nc__LacI_P1 </ci>
114             </apply>
115           </apply>
116         </apply>
117       </math>
118       <listOfLocalParameters>
119         <localParameter id="RNAP" units="u_1_mole_n1" value="30"/>
120         <localParameter id="ng__P1" units="u_1_mole_n1" value="4"/>
121         <localParameter id="Ko__P1" units="u_1_mole_n1" value="0.033"/>
122         <localParameter id="ko__P1" units="u_1_second_n1" value="0.05"/>
123         <localParameter id="nc__LacI_P1" units="dimensionless" value="2"/>
124         <localParameter id="Kr__LacI_P1" units="u_1_mole_n1" value="0.5"/>
125       </listOfLocalParameters>
126     </kineticLaw>
127   </reaction>
128   • • • •
```

Fig. 9.11 Cropped image of data stored in the *List of Reactions* variable

program jumps directly to the next <*reaction* tag, with the help of shift register, thus by passing all the intermediate local parameter *id*=" " fields.

Now we need to extract the list of reactants, products, and modifiers. It is worth mentioning here that a reaction may have no reactant, or a product, or a modifier species, or it may contain all of them. For example, a degradation reaction of any

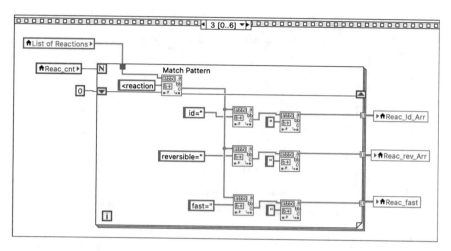

Fig. 9.12 Extracting the reactions IDs and other basic attributes

specific species would only have a reactant which will degrade over time and it would not produce any product. Similarly, the reaction data shown in Fig. 9.11 indicates that there are no reactants involved in the reaction, rather a modifier, *LacI* (line 76), controls the production of the product species *CI* (line 73). Therefore, before looking for the reactants, products, and modifiers, we first need to find out if the reaction has the *<listOfReactants>*, *<listOfProducts>*, *<listOfModifiers>*, and the *<listOfLocalParameters>*.

 Modifier is the species which is not a reactant in a reaction but has a direct impact on the production of reaction's product. It can be considered as a Catalyst which may or may not degrade over time.

Figure 9.13 indicates the frame number 4 of the *stacked sequence structure* in our program. This figure indicates that, for every reaction in the model, all of the above mentioned four lists are processed in parallel in separate nested sequence structures. First, in the 0th frame of all the nested sequence structures, the program determines the presence of the above mentioned lists. The results of the presence of the *<listOfReactants>*, *<listOfProducts>*, *<listOfModifiers>*, and the *<listOfLocalParameters>* are stored in the Boolean variables named *R-lst*, *P-lst*, *M-lst*, and *LP*, respectively.

Figure 9.14 shows the first frame of all nested sequence structures inside the *for* loop. To make the code clearly visible, only the cropped images of these frames are included in this figure. It can be noticed that the program executes ahead only when the corresponding list exists in the SBML model, which is controlled by the *case structures*. When the corresponding list does not exist, the left side terminal of a *shift register* directly connects to its right side terminal through the *False* frame of *case structure* (not shown in the Figure). When the corresponding list exists, it can be seen that the program executes ahead in sequence, using another nested **stacked**

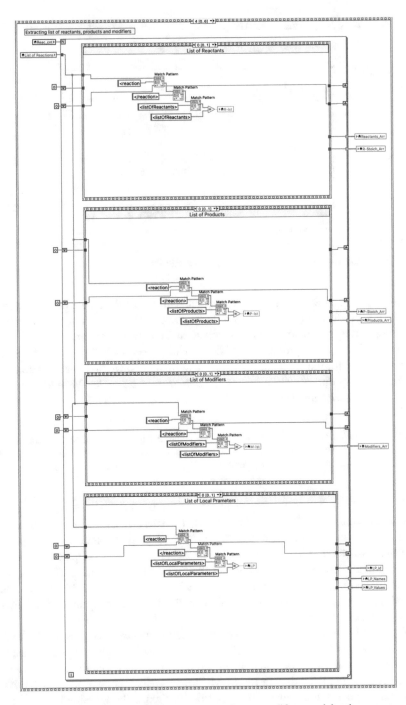

Fig. 9.13 Determining if the lists of reactants, products, modifiers, and local parameters are present in each reaction of the SBML model

sequence structure, and first fetch the contents of the corresponding lists (shown in frame 0[0..4] in Fig. 9.14).

 From Fig. 9.14 until 9.16, only the nested code will be displayed. Note that the figures presented ahead are further nested inside the outer frames (1[0..1]) shown in Fig. 9.14, and all of which are enclosed inside the *for* loop as shown in Fig. 9.13.

In the next sub-sequences (1 [0..4]), the number of corresponding attributes, for example, *reactants*, is determined using code similar to that used to extract the number of species, as shown in Fig. 9.8. The token operator and token strings used for counting the reactants, products, modifiers, and local parameters are *<species-Reference*, *<speciesReference* (again for products), *<modifierSpeciesReference*, and *<localParameter*, respectively. The frames for these codes are not shown separately as they are the exact replication of Fig. 9.8 with the above mentioned tokens replaced.

In this way, we can determine the number of reactants, products, modifiers, and local parameters involved in the reaction being processed in the current iteration of the *for* loop. The values of the number of reactants, products, modifiers, and local parameters are stored in the *Rctns_Indx*, *Prdct_Indx*, *Mod_Indx*, and *LP_Indx* numeric variables, respectively.

Once the number of corresponding attributes is determined (in frame 1), we can extract and store them in separate arrays. A reaction may have multiple reactants, products, modifiers, and local parameters. We extract all such multiple attributes one-by-one using a *for* loop, and store them as comma-separated values in the same variable. The combined value of the variable is then inserted into an array to hold the attributes of all reactions as a separate array element. For example, Fig. 9.15 shows the second frame (2 [0..4]) of nested sequence structure for the list of products where the product species and their corresponding stoichiometries are extracted. In our current example of the reaction, there is only one product species, *CI*, with a stoichiometric value of *10*, as shown in line 73 in Fig. 9.11.

A separate case structure is used, each for species and stoichiometry, to store the attributes as comma-separated values in the separate variables, named *Products* and *P-Stoich*, respectively. The first attribute value is stored directly in the respective variables (via the default case structure for $i = 0$, not shown). The new values, for $i >= 1$, are concatenated with the previous values separated by a comma, as shown in Fig. 9.15.

The exact code (Fig. 9.15) will be used to extract the list of reactants and their corresponding stoichiometries, except that the values will be stored in separate variables, named *Reactants* and *R-Stoich*, respectively. Also, *for* loop will run for the number of times specified by the *Rctns_Indx* value. Similarly, the attributes of modifiers (species only) and local parameters (id, name and value) can be obtained.

 Implement the logic to obtain and store the name of Modifier species in *Modifiers* variable. Also, obtain and store the id, name, and value of local parameters, and store their values in *LP_ID*, *LP_Names*, and *LP_Values* variables, respectively.

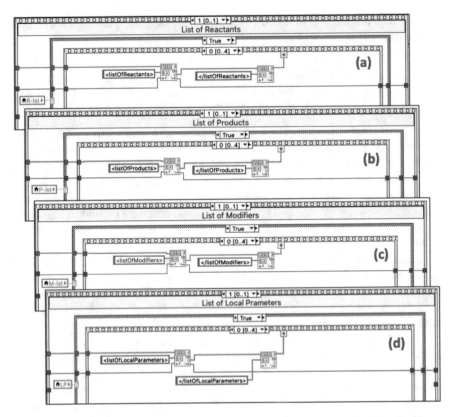

Fig. 9.14 Frame 1 of all nested sequence structures for: (**a**) List of reactants. (**b**) List of products. (**c**) List of modifiers. (**d**) List of local parameters

Once all the attributes related to all the four lists, for the specific reaction, are obtained, we can insert them in the separate arrays and reset the variables to hold the values of next reaction in the next iteration of outer *for* loop. Figure 9.16a and (b) shows that the values are first stored in *Reactants_Arr* and *R-Stoich-Arr* arrays (frame 3), and then the variables are reset (frame 4), respectively.

Similarly, implement frames 3 and 4 for the list of products, list of modifiers, and the list of local parameters and store their values in the arrays as named below:

- Products –> Products_Arr
- P-Stoich –> P-Stoich_Arr
- Modifiers –> Modifiers_Arr
- LP_ID –> LP_ID_Arr
- LP_Name –> LP_Names_Arr
- LP_Value –> LP_Values_Arr

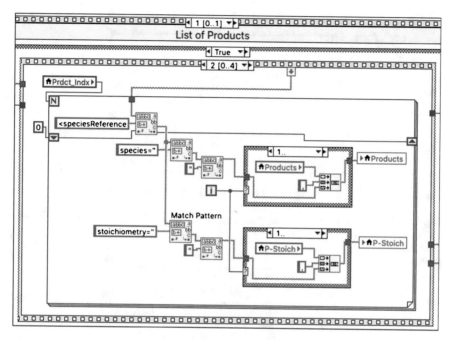

Fig. 9.15 Extraction of product species and the corresponding stoichiometries

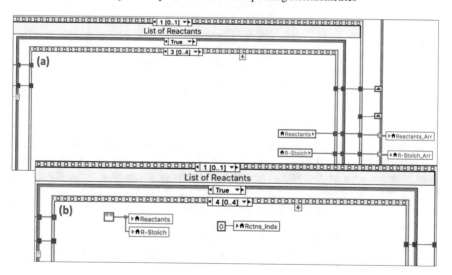

Fig. 9.16 Last two frames of sub-sub sequence structure. (**a**) Storing the values in separate arrays. (**b**) Resetting the variables

Frame 4 of the main sequence structure, shown in Fig. 9.13, is now completed and the necessary attributes of all reactions are obtained, except the kinetic law, which are extracted in the next frame as shown in Fig. 9.17.

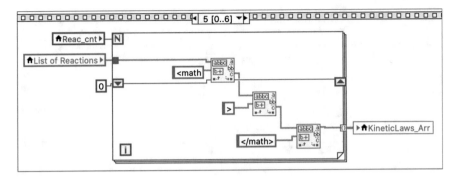

Fig. 9.17 Extraction of kinetic laws and storing in an array named *KineticLaws_Arr*

Fig. 9.18 Creation of single multi-dimensional array to hold all the attributes of reactions

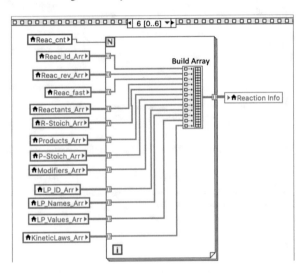

In SBML standard, the kinetic laws are represented in the *MathML* (Mathematical Markup Language) format. MathML is cast as an application of XML and is used to mark up a mathematical equation in terms of presentation and semantics. For example, the MathML of the kinetic law (Fig. 9.11) enclosed in *<math>* tags (lines 79–117) represents the following equation:

$$\frac{ko\_\_P1 * ng\_\_P1 * Ko\_\_P1 * RNAP}{1 + Ko\_\_P1 * RNAP + (Kr\_\_LacI\_P1 * LacI)^{nc\_\_LacI\_P1}} \tag{9.1}$$

Finally, we gather all the arrays holding different attributes of reactions in a single multi-dimensional array, named *Reactions_Info*, as shown in Fig. 9.18.

 As same as you sorted the species data in order before, do the same for arranging the *Reaction Info* in ascending order according to their IDs.

As same as we created the subVIs for *Species Parser* and *Parameter Parser* in previous sections, we have to create a subVI for *Reaction Info Parser* and use it in the main VI as shown in Fig. 9.5b. Now assign the inputs (*Full SBML File*) and the outputs (*Reac_Cnt* and *Reactions_Info*) in the *connector pane* (see Fig. 7.15 in Sect. 7.3.4). Also, refer to Sect. 7.3.4.1 to create an icon for this sub VI such that it matches to the one shown in Fig. 9.5c.

Figure 9.5c further shows that the array of *Kinetic Laws* has been extracted and used in the next frame (5 [0..5]) to convert the MathML expressions into a readable form, similar to Eq. 9.1.

9.2 Challenge

In this challenge, you are required to implement the last, yet the most challenging, part of the SBML file parsing. That is, you are required to obtain the MathML expressions in a standard mathematical representation.

Figure 9.5d gives a hint that it can be implemented into two steps—first converting the MathML expression into an intermediate expression, using a subVI named *Process Kinetic Law* and then transform that intermediate expression into the desired form, using another sub VI named *Math Finder*.

As can be seen in Fig. 9.11, the entire expression is enclosed in *$-$* tags containing multiple sub expressions enclosed in *<apply>-</apply>* tags. Now the challenge is to identify the matching *<apply>-</apply>* tags which are nested within each other. For example, look at the lines 80–95 (in Fig. 9.11) which are reproduced in Fig. 9.19 for convenience. It can be noticed in this figure that the first *<apply>* tag starts at line 80, which includes more *<apply>* tags at lines 82, 84, and 89. The *<apply>* tags which start at line 82, 84, and 89 are closed at lines 88, 93, and 94, respectively.

Following list shows the expressions composed from the respective *<apply>* tags, shown in Fig. 9.19. The lines 82–94 enclose two multiplication expressions inside it, i.e., lines 84–88 and 89–93. The entire expression between lines 82–94 constitutes the numerator of Eq. 9.1.

- Lines 84–88: $ko\_\_P1 * ng\_\_P1$
- Lines 89–93: $Ko\_\_P1 * RNAP$
- Lines 82–94: $ko\_\_P1 * ng\_\_P1 * Ko\_\_P1 * RNAP$

To retrieve the expression programmatically, it is tricky to match the closing *</apply>* tag with its corresponding opening *<apply>* tag. For instance, when a program crawls through *<apply>* tags, in Fig. 9.19, and encounters the first closing *</apply>* tag at line 88, the closing one should be associated with the corresponding *<apply>* tag at line 84 and not with those present at lines 82 and 80.

HINT This problem can be solved by assigning unique opening and closing tags to *<apply>* and *</apply>* tags. For example, the subVI, *Process Kinetic Law*, shown

Fig. 9.19 Example of nested
<apply> tags

```
                                           ⋮
   80               <apply>
   81                   <divide/>
   82                   <apply>
   83                       <times/>
   84                       <apply>
   85                           <times/>
   86                           <ci> ko__P1 </ci>
   87                           <ci> ng__P1 </ci>
   88                       </apply>
   89                       <apply>
   90                           <times/>
   91                           <ci> Ko__P1 </ci>
   92                           <ci> RNAP </ci>
   93                       </apply>
   94                   </apply>
   95                   <apply>
                                           ⋮
```

in Fig. 9.5d, assigns *SOP<X>* tags to opening *<apply>* tags and the *EOP<X>* to
the closing *</apply>* tags, where *<X>* corresponds to a numeric digit. You may
also consider replacing other mathematical operator tags with their corresponding
standard math operators.

To explain the above concept, let us consider the portion of MathML shown in
Fig. 9.19. When the program encounters the first *<apply>* tag at line 80, it would
replace the tag with unique identifier tag *SOP1*. Next the *<divide/>* operator tag
can be replaced with the division operator "/". Proceeding ahead, the next *<apply>*
tag would be replaced with a higher value tag, i.e., *SOP2*. Then comes the next
operator *<times/>* tag which can be replaced with its corresponding mathematical
operator, i.e., "*". Similarly the next higher *<apply>* tag, at line 84 would be
assigned a higher value (*SOP3*) and the next mathematical operator is replaced with
its corresponding actual operator. The *<ci>- </ci>* are content identifier tags that are
used to hold the variables. These variables are commonly the names of parameters
or local parameters defined in the kinetic law for any reaction. The name of the
variable enclosed in content identifier tags can simply be extracted as it is.

Now the first closing *</apply>* tag encounters at line 88, which has to be replaced
with the *EOP* tag along with a numeric digit associated with it. Here, the program
can see the current counter value of *SOP* tag and associate it with the current EOP
tag. For instance, in this example, the current value of *SOP* tag is 3, therefore the
closing *</apply>* tag can be replaced with *EOP3*. Once the matching *EOP* tag is
encountered, the SOP counter is decreased by 1 to indicate that the pair of SOP3 and

Fig. 9.20 Processed
expression with unique tags

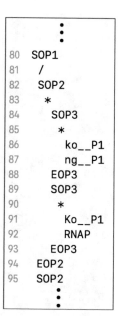

EOP3 is completely accessed. The next *<apply>* tag at line 89, would be given the *SOP3* tag again to indicate another nested expression within the *SOP2* tag. This may help a program to clearly distinguish the nested expressions using these unique tags. The *processed* expression, corresponding to the one shown in Fig. 9.19, is shown in Fig. 9.20.

Once the expression is processed, the next subVI, *Math Finder*, converts the expression, into a standard mathematical form, similar to the one shown in Eq. 9.1.

The logic you develop in *Math Finder* may need to be called recursively to transform the processed expression in to standard mathematical representation.

The front panel of the final SBML parser would look like the one shown in Fig. 9.21, containing the information about species, parameters, reactions, and math expressions of kinetic laws.

9.3 Summary

In this chapter, we have practically experienced how to parse SBML file using LabVIEW. We developed a parser for some important SBML components, though being optional, but are usually present in every model of a genetic circuit. The same approach can be followed to parse other SBML components.

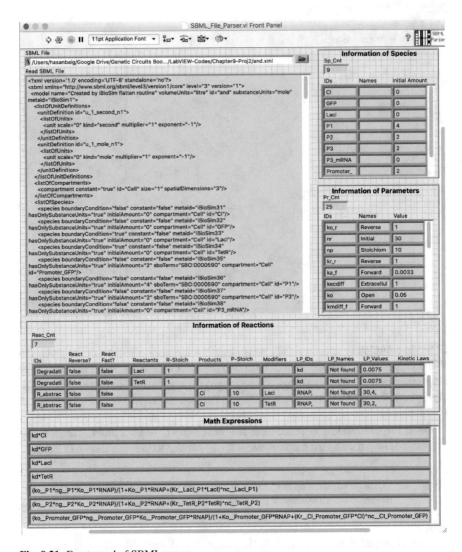

Fig. 9.21 Front panel of SBML parser

Another interesting project would be to develop the entire G-library for reading and parsing the SBML file, and make it available for public use like other SBML libraries including jSBML, libSBML, etc.

Index

© Springer Nature Switzerland AG 2020
H. Baig, J. Madsen, *Genetic Design Automation*,
https://doi.org/10.1007/978-3-030-52355-8

Printed in the United States
by Baker & Taylor Publisher Services